Oral Microbial Communities and Oral Health

Oral Microbial Communities and Oral Health

Editor

Yoshiaki Nomura

MDPI • Basel • Beijing • Wuhan • Barcelona • Belgrade • Manchester • Tokyo • Cluj • Tianjin

Editor
Yoshiaki Nomura
Tsurumi University School of
Dental Medicine
Japan

Editorial Office
MDPI
St. Alban-Anlage 66
4052 Basel, Switzerland

This is a reprint of articles from the Special Issue published online in the open access journal *Applied Sciences* (ISSN 2076-3417) (available at: https://www.mdpi.com/journal/applsci/special_issues/human_oral_microbiome).

For citation purposes, cite each article independently as indicated on the article page online and as indicated below:

LastName, A.A.; LastName, B.B.; LastName, C.C. Article Title. *Journal Name* **Year**, *Volume Number*, Page Range.

ISBN 978-3-0365-2880-9 (Hbk)
ISBN 978-3-0365-2881-6 (PDF)

Contents

About the Editor

Yoshiaki Nomura, DDS, PhD, was Professor of Department of Translational Research, Tsurumi University School of Dental Medicine, when editing this book. He is professor of International Photocatalyst Research Institute, University of Shanghai for Science and Technology, He graduated from Tokyo Medical and Dental University, then went on to complete his doctorate at the Medical Graduate Course of Medical Research Institute of Tokyo Medical and Dental University. He joined the department of molecular and cell biology at the medical research institute of Tokyo Medical and Dental University and Department of Biochemistry and Cell Biology, National Institute of Infectious diseases, Japan. He worked as a lecturer at the Department of Preventive dentistry and public health, then, Director of Section of Oral Health Technology, Department of Oral Health National Institute of Public Health, Japan. He specializes in statistical modeling, molecular biology, and bioinformatics. He has more than 130 publications in peer-reviewed international journals (h-index 31).

Preface to "Oral Microbial Communities and Oral Health"

The oral microbiome has a great impact on human health. Next-generation sequencing and bioinformatics provide us valuable information on the role of oral health in overall heath and diseases. The research field has propagated from infectious diseases to non-communicable disease. The oral microbiome changes throughout the life stage, with an effect on dental diseases. Thus, we need to identify the pathogenic oral microbiome, not only for oral disease but also for systemic diseases. However, at this stage, the information regarding healthy or pathogenic oral microbiome has not been clearly identified. In this book, information regarding the current knowledge on and advancement of research into oral microbiomes is covered. The subjects covered range from children to centenarian. The diseases covered are stain, dental caries, periodontal disease, tooth malformation, and systemic disease. The information in this book may be useful, not only for researchers and clinicians of oral health, but a wide range of health professionals as well as life science researchers.

Yoshiaki Nomura
Editor

Editorial

Future Prospective of Oral Microbiome Research

Yoshiaki Nomura [1,2,*], Ayako Okada [3] and Nobuhiro Hanada [2]

[1] Department of Translational Research, Tsurumi University School of Dental Medicine, Yokohama 230-8501, Japan
[2] International Photocatalyst Research Institute, University of Shanghai for Science and Technology, Shanghai 200093, China; gtfgene@gmail.com
[3] Department of Operative Dentistry, Tsurumi University School of Dental Medicine, Yokohama 230-8501, Japan; okada-a@tsurumi-u.ac.jp
* Correspondence: nomura-y@sirius.ocn.ne.jp or nomura-y@tsurumi-u.ac.jp

Abstract: Oral microbiome has complex structure. It consisted of more than 700 species of bacteria. These bacteria contains pathogens for human health. In contrast, some beneficial bacteria were included. Perspective of oral microbiome is not still elucidated. In this paper, information regarding oral microbiome of health older adults and oral diseases are included. Additionally, concise review of oral microbiome are presented.

Keywords: oral microbiome; next generation sequence; clinical application

Citation: Nomura, Y.; Okada, A.; Hanada, N. Future Prospective of Oral Microbiome Research. *Appl. Sci.* **2022**, *12*, 55. https://doi.org/10.3390/app12010055

Received: 16 December 2021
Accepted: 21 December 2021
Published: 22 December 2021

Publisher's Note: MDPI stays neutral with regard to jurisdictional claims in published maps and institutional affiliations.

An oral microbiome consists of more than 700 species of bacteria [1] and plays an important role in human health. In the past, disease-specific bacteria were focused on, many in vitro studies showed their pathogenesis, and their clinical examination system has developed. However, these disease-specific bacteria configure small, tiny fractions of oral microbiomes. The advance of next-generation sequencing and bioinformatics enabled us to study the proportion of each species in samples [2]. Despite this, there is still no clear definition of a healthy oral microbiome. The disturbance of a healthy oral microbiome, which is called dysbiosis, immensely contributes to human health. Therefore, there is a need to accumulate the data on the oral microbiome for every age category in humans, in addition to the disease stage.

In this article, the key species proposed in this paper are summarized in Table 1. Many of these species are relatively unfamiliar as the etiology of dental caries, periodontal disease, or other oral diseases. The pathogen of black stain is almost unknown. This study may give us the candidate for the pathogens of black stain for future research [3]. Periodontal pathogens in patients with Down syndrome were suggested to be different from chronic marginal periodontitis [4]. Molar–incisor malformation is a very rare disease, and the case report by [5] may contain valuable information regarding this. In general, *Firmicutes* is a predominant phylum, and *Streptococcus* is a predominant genius. These bacteria are supposed to be beneficial bacteria for human health. Some of them had been used as probiotic bacteria. However, some pathogenic species are predominant in healthy subjects. For example, predominant bacteria such as the *S. sinensis* group and *S. pneumoniae* group are known to be the pathogens for pneumonia [6,7].

In addition to this clinical evidence, this paper introduces the effect of biofilm composition on an experimental animal model for the development of dental caries [8] and reviews the effect of silver diamine fluoride on biofilm composition [9].

As the cost of next generation sequencing is high for the sample size study, the experiments presented in this paper were carried out by a small sample size study. Therefore, there is a need to accumulate the data of oral microbiome. The clinical application of the microbiome and its future prospects are reviewed [10]. Moreover, we hope that this paper may help guide future studies.

Table 1. List of the key species proposed in this paper.

Genus	Species	Method
Predominant in patients with black stain [3]		
Abiotrophia	*Ab. defectiva*	
Actinomyces	*Ac. naeslundii*	
Cardiobacterium	*C. hominis*	
Eikenella	*E. corrodens*	
Granulicatella	*G. adiacens*	
Kingella	*JQ455165_s*	
Leptotrichia	*L. hongkongensis*	
Neisseria	*N. elongata*	Next generation sequence
	N. perflava	
	N. sicca group	
	N. subflava	
Porphyromonas	*AM420091_s*	
	P. catoniae	
	4P003207_s	
Selenomonas	*Se. noxia*	
Streptococcus	*St. sanguinis*	
Periodontal disease in patients with Down Syndrome [4]		
Eikenel	*E. corrodens*	qPCR
Tannerella	*T. forsythia*	
Patients with molar–incisor malformation (case report) [5]		
Predominant genera		
Streptococcus	-	
Veilonella	-	Next generation sequence
Leptotrichia	-	
Severe periodontal abscess:		
Spirochetes	-	
Core oral microbiome of female centenarians [6]		
Streptococcus	*S. salivarius group*	
	S. sinensis group	
	S. pneumoniae group	
	S. parasanguinis group	
	S. gordonii group	
Veillonella	*V. dispar*	Next generation sequence
	V. parvula group	
Granulicatella	*G. adiacens group*	
Lactobacillus	*L. salivarius*	
Rothia	*R. mucilaginosa*	
	R. dentocariosa	
Prevotella	*P. histicola*	

Table 1. *Cont.*

Genus	Species	Method
Core oral microbiome of healthy older adult [7]		
Atopobium	*A. parvulum*	
Actinomyces	*KE952139_s*	
Streptococcus	*S. sinensis*	Next generation sequence
	S. pneumoniae	
	S. salivarius	
	S. peroris	
	S. parasanguinis	
	Streptococcus_uc	
	AFQU_s	
Granulicatella	*G. adiacens*	
Rothia	*KV831974_s*	
Veillonella	*V. dispar*	
	V. parvula	

Author Contributions: Conceptualization, Y.N., A.O. and N.H.; writing—original draft preparation, Y.N.; writing—review and editing, A.O. and N.H. All authors have read and agreed to the published version of the manuscript.

Funding: This study was supported by the JSPS KAKENHI (grant numbers 17K12030 and 20K10303) and the SECOM Science and Technology Foundation. The funders played no role in the design of the study, in data collection, in the analysis and interpretation of the results, or in the writing of the manuscript.

Institutional Review Board Statement: Not applicable.

Informed Consent Statement: Not applicable.

Data Availability Statement: Not applicable.

Conflicts of Interest: The authors declare no conflict of interest.

References

1. Aas, J.A.; Paster, B.J.; Stokes, L.N.; Olsen, I.; Dewhirst, F.E. Defining the normal bacterial flora of the oral cavity. *J. Clin. Microbiol.* **2005**, *43*, 5721–5732. [CrossRef] [PubMed]
2. Keijser, B.J.; Zaura, E.; Huse, S.M.; van der Vossen, J.M.; Schuren, F.H.; Montijn, R.C.; ten Cate, J.M.; Crielaard, W. Pyrosequencing analysis of the oral microflora of healthy adults. *J. Dent. Res.* **2008**, *87*, 1016–1020. [CrossRef]
3. Hwang, J.; Lee, H.; Choi, J.; Nam, O.; Kim, M.; Choi, S. The Oral Microbiome in Children with Black Stained Tooth. *Appl. Sci.* **2020**, *10*, 8054. [CrossRef]
4. Cuenca, M.; Marín, M.; Nóvoa, L.; O'Connor, A.; Sánchez, M.; Blanco, J.; Limeres, J.; Sanz, M.; Diz, P.; Herrera, D. Periodontal Condition and Subgingival Microbiota Characterization in Subjects with Down Syndrome. *Appl. Sci.* **2021**, *11*, 778. [CrossRef]
5. Lee, H.; Kim, H.; Lee, K.; Kim, M.; Nam, O.; Choi, S. Complications of Teeth Affected by Molar-Incisor Malformation and Pathogenesis According to Microbiome Analysis. *Appl. Sci.* **2021**, *11*, 4. [CrossRef]
6. Nomura, Y.; Kakuta, E.; Okada, A.; Otsuka, R.; Shimada, M.; Tomizawa, Y.; Taguchi, C.; Arikawa, K.; Daikoku, H.; Sato, T.; et al. Oral Microbiome in Four Female Centenarians. *Appl. Sci.* **2020**, *10*, 5312. [CrossRef]
7. Nomura, Y.; Kakuta, E.; Kaneko, N.; Nohno, K.; Yoshihara, A.; Hanada, N. The Oral Microbiome of Healthy Japanese People at the Age of 90. *Appl. Sci.* **2020**, *10*, 6450. [CrossRef]
8. Chen, J.; Kong, L.; Peng, X.; Chen, Y.; Ren, B.; Li, M.; Li, J.; Zhou, X.; Cheng, L. Core Microbiota Promotes the Development of Dental Caries. *Appl. Sci.* **2021**, *11*, 3638. [CrossRef]
9. Zhang, J.; Got, S.; Yin, I.; Lo, E.; Chu, C. A Concise Review of Silver Diamine Fluoride on Oral Biofilm. *Appl. Sci.* **2021**, *11*, 3232. [CrossRef]
10. Cho, Y.; Kim, K.; Lee, Y.; Ku, Y.; Seol, Y. Oral Microbiome and Host Health: Review on Current Advances in Genome-Wide Analysis. *Appl. Sci.* **2021**, *11*, 4050. [CrossRef]

Review

Oral Microbiome and Host Health: Review on Current Advances in Genome-Wide Analysis

Young-Dan Cho, Kyoung-Hwa Kim, Yong-Moo Lee, Young Ku and Yang-Jo Seol *

Department of Periodontology, School of Dentistry and Dental Research Institute, Seoul National University and Seoul National University Dental Hospital, Seoul 03080, Korea; cacodm1@snu.ac.kr (Y.-D.C.); perilab@snu.ac.kr (K.-H.K.); ymlee@snu.ac.kr (Y.-M.L.); guy@snu.ac.kr (Y.K.)
* Correspondence: yjseol@snu.ac.kr; Tel.: +82-2-2072-0308

Abstract: The oral microbiome is an important part of the human microbiome. The oral cavity has the second largest microbiota after the intestines, and its open structure creates a special environment. With the development of technology such as next-generation sequencing and bioinformatics, extensive in-depth microbiome studies have become possible. They can also be applied in the clinical field in terms of diagnosis and treatment. Many microbiome studies have been performed on oral and systemic diseases, showing a close association between the two. Understanding the oral microbiome and host interaction is expected to provide future directions to explore the functional and metabolic changes in diseases, and to uncover the molecular mechanisms for drug development and treatment that facilitate personalized medicine. The aim of this review was to provide comprehension regarding research trends in oral microbiome studies and establish the link between oral microbiomes and systemic diseases based on the latest technique of genome-wide analysis.

Keywords: oral microbiome; oral disease; systemic disease; genome-wide analysis; personalized medicine

Citation: Cho, Y.-D.; Kim, K.-H.; Lee, Y.-M.; Ku, Y.; Seol, Y.-J. Oral Microbiome and Host Health: Review on Current Advances in Genome-Wide Analysis. *Appl. Sci.* **2021**, *11*, 4050. https://doi.org/10.3390/app11094050

Academic Editor: Yoshiaki Nomura

Received: 30 March 2021
Accepted: 27 April 2021
Published: 29 April 2021

Publisher's Note: MDPI stays neutral with regard to jurisdictional claims in published maps and institutional affiliations.

1. Introduction

The human oral cavity is the second largest microbial habitat comprising fungi, bacteria, and viruses, after the intestines, and has a special niche consisting of soft tissue of the gingiva and oral mucosa and the hard tissue of the teeth [1]. Although the oral cavity and intestines are a distance apart, they are connected and show a marvelous diversity of microorganisms. The oral microbiome refers to the collective genetic materials of oral microorganisms. The microorganisms in the oral cavity are referred to as oral microbiota, oral microflora, and oral microbiome [2]. Microorganisms coexist in our bodies and live through symbiosis, dysbiosis, and pathogenic relations. The microbiome not only refers to the microorganisms involved, but also includes areas of activity that result in the formation of specific ecological niches. Significant changes in the local environment can disrupt the microbe–microbe interaction, which can alter the host–microbe equilibrium, increasing the risk of disease [3]. The oral cavity is the most accessible habitat for studying the relationship between the host and microorganisms. Different oral structures such as the teeth, the gingiva, the palate, the cheek, and the lips are colonized by distinct microbial communities, and the surfaces of the oral cavity are covered with bacterial biofilm [4]. The special environment of the oral cavity, with its stable pH of 6.5–7.0 of saliva, moisture, and its temperature of an average of 37 °C, creates the favorable conditions necessary for the growth of microorganisms [5]. For this reason, various kinds of microorganisms are well distributed in the mouth. The oral cavity is one of the best-studied microbiomes to date, with research focusing on its role as a part of human microbiome development and how it influences systemic health and disease. About 700 species of prokaryotes have been identified [6], and 392 taxa with at least one reference genome exist [7]. In the near future, the advances in our understanding of oral microbiomes may change our way of life through in-depth principles of oral biology and novel therapeutics. In this review, we provide a general understanding of the oral microbiome, establish the link between oral microbiomes and general health based on a genome-wide analysis through the latest technique, and discuss the perspective and future directions for both oral and systemic health.

2. Paradigm Shift: Microorganisms to Microbiomes

The term "microorganism" simply refers to very small living things such as archaea, bacteria, protists, fungi, and viruses, while "microbiome" is a collective and comprehensive term for microorganisms. The community of microbial residents is referred to as the microbiome, and recent research on microbiomes has demonstrated an important role of the communities of microorganisms in human homeostasis (Figure 1A) [8]. The term "microbiome" was coined by Joshua Lederberg, a Nobel Prize laureate, "to signify the ecological community of commensal, symbiotic, and pathogenic microorganisms that literally share our body space and have been all but ignored as determinants of health and disease" [9]. As a collection of information, the microbiome includes the microorganisms' genomic data, structural elements, metabolites, and environmental conditions (Figure 1A) [10]. With the emergence of new technology, including next-generation sequencing (NGS), and in-depth information, such as sequence profiles of microbial communities, numerous insights on the relation between the human microbiome and disease have been obtained [11].

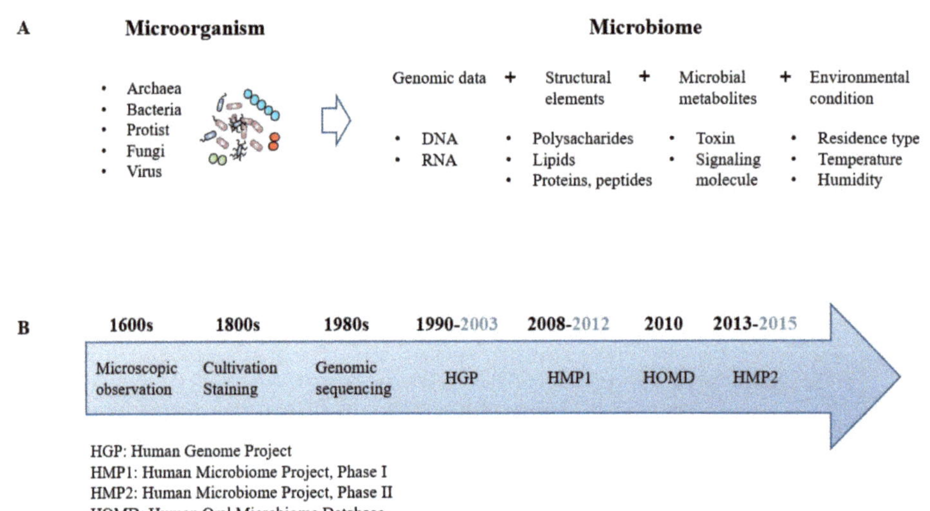

Figure 1. From microorganism to microbiome; a paradigm shift with advanced technology. (**A**) The term, microbiome, refers not only to the microorganisms but also to their theatre of activity. (**B**) Timeline of microbiome research.

In the 1670s, the study of human microbiology was initiated with the invention of the microscope, and the consequent observation of bacteria in pond water and dental plaque, by Antoni Philips van Leeuwenhoek [12,13] (Figure 1B). He was the first to observe microorganisms and determine their sizes. After that revolutionary invention, a culture-based method that is dependent on the growth of viable and culturable microbes was introduced to identify microbes with biochemical subtyping in the 1800s [14]. After the 1980s, the emergence of new technology, such as DNA sequencing, which enables acquisition of more than 1 trillion sequences of genetic information, provided a powerful unculturable method for understanding human health and disease [15]. Based on this advance in sequencing technology, the Human Genome Project (HGP) was started in 1990 and completed in 2003. The first sequence of the human genome was published in 2001 and constituted about 90% of the genome [16]. With the HGP, the elucidation of the human genome initiated a new era in "precision medicine", which is the ideal concept of health care [17]. The basis of precision medicine includes the genomes of humans and of the microorganisms, such as fungi, viruses, and bacteria, that reside in the human body. As the extension of the HGP, the Human Microbiome Project (HMP) was launched in 2007 with the goal of establishing 3000 microbial genomes from the representative six regions of the body (the mouth, the esophagus, the stomach, the colon, the vagina, and the skin) as reference genomes [18].

3. Methodology of the Microbiome Research

Completion of the HGP was accompanied by parallel and ongoing development of commercially available high-throughput DNA sequencing tools, which eventually facilitated the acquisition of genomic information from human samples. High-throughput techniques such as NGS have replaced the culture-based traditional methods for identifying and characterizing microbes [15]. Past studies without NGS and relying on the laboratory culture system were limited to small parts, such as the biochemical or physiological properties of microbiomes, called opportunistic pathogens [19]. In the beginning, microbiology research had focused on identifying pathogens within the commensal microbiota and deciphering their virulence in relation to health and disease. In the 1980s, a sequencing technique was introduced that enabled nucleotide sequences to be determined. This remarkable advance provided new and diverse information about microbial taxonomic profiling regardless of cultivability [19]. In taxonomic profiling, two methods, amplicon sequencing (16S rRNA sequencing) and direct shotgun sequencing (metagenomics), are used [19]. In the oral microbiome studies, samples were taken from dental plaque and crevicular fluid from the periodontal pocket. This technology has allowed the characterization of microbial diversity in the human microbiome with unprecedented depth and coverage [20]. These oral microbiome studies originated from HMP's 16S rRNA sequencing data, published in the Human Oral Microbiome Database (HOMD), which identified oral bacteria from specific locations in the mouth such as the teeth, the gingival sulcus, the tongue, the cheek, the tonsils, and the soft and hard palates [2]. These early sequencing studies focused on the 16S rRNA sequence, and this conventional sequencing method was performed using a cultivated clonal culture; however, there were many microorganisms that were unculturable and thus could not be identified with this conventional method (Figure 2) [21]. This cultivation-based method could find < 1% of microbial species; therefore, metagenomics was introduced to identify non-isolated organisms via DNA sequencing without the need for cultivation and isolation [21]. In addition to metagenomics, DNA barcoding or metabarcoding is also used for the identification of species, using a short section of DNA from a reference library database [22]. The term "metabarcoding" is used when DNA barcoding is used to identify organisms from samples containing DNA from more than one organism [23]. Metagenomics also enables the metabolic and functional diversity of microbial communities to be accessed via metatranscriptomics, metaproteomics, and metabolomics [24]. Metaproteomics has emerged as a complementary approach to identify the functions of microbial communities [25]. Mass spectrometry-based proteomics has been widely used for studying the composition of proteins in oral microbial communities and has been applied to characterize oral biofilms [26]. Many metaproteomic studies have been performed in periodontal disease and endodontic infections associated with apical periodontitis using saliva and crevicular fluids [27]. Metabolomics is the study of metabolites within a biologic system to discover what happens in cells. Several studies have been conducted on the salivary metabolome; however, these have offered less specific explanations due to difficulties regarding disentangling host reactions and microbial contributions [27]. As such, metaproteomics and metabolomics complement the foundations established by metagenomics and metatranscriptomics in the comprehensive understanding of microbiomes.

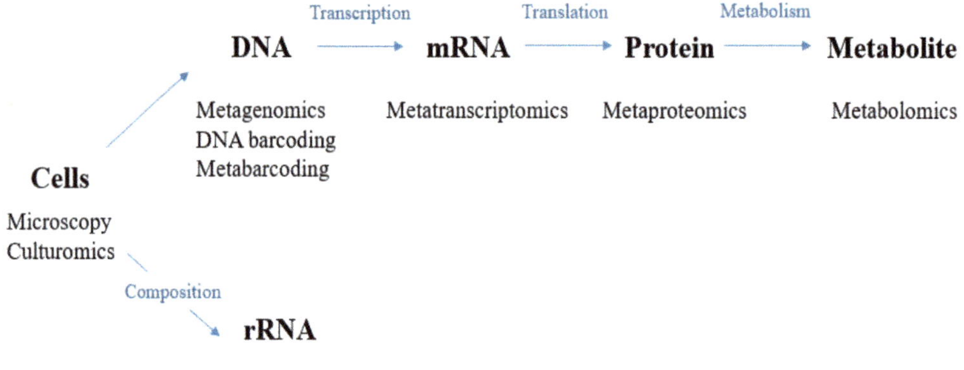

Figure 2. Methodology of the microbiome research.

4. Core, Pathologic, and Healthy Microbiomes

Microbiomes are largely divided into two types: core and variable microbiomes. The core microbiome is common to all individuals, while the variable microbiome is unique to individuals according to their lifestyle [28]. One of the goals of the HMP was to identify "core microbiomes" defined as microbial taxa or genes shared by most people [29]. The core microbiomes are described via five types: common core, temporal core, ecological core, functional core, and host-adapted core [28]. In order to diagnose and treat disease at an early stage, it is necessary to explain the commensal microbiome associated with health [30]. The commensal human microbiome is estimated to be 10 times greater in number than the quantity of human cells. These microbial communities are normal residents of the oral cavity, the skin, and the intestinal mucosa, and have a wide range of functions that are essential to the survival of the host. When the symbiotic balance between the host and the microbes is disrupted and disease is obvious, the presence of pathologic microbes stands out [31]. On the other hand, if the healthy microbiome is dominated by a "core microbiome", the homeostasis of health would be maintained [32]. Zaura et al. defined the healthy core microbiome of oral microbial communities from several intraoral niches using 454 pyrosequencing, showing a major proportion of oral bacterial sequences of healthy individuals as identical, which supports the concept that there are key microbiomes in health [30]. In addition, Bao et al., identified the microbiome and proteomic profiles in the gingival tissue of healthy individuals and individuals with periodontitis by a pressure cycling technology-assisted workflow, which is an emerging platform for tissue homogenization and sequence retrieval coverage [33]. The results showed that 69 proteins were differentially expressed in periodontitis, and Treponema sp. HMT253 and Fusobacterium naviforme were strongly associated with disease sites, indicating the existence of a tissue-specific microbiome signature.

5. Human Microbiome Project

The omics era, with its innovative sequencing techniques, accelerated all aspects of biological research, and its effects were particularly evident in studies on microbial communities and the human microbiome. Genomic sequencing has revealed the diversity of microbial communities in the body, and more information has been obtained as the National Institute of Health Human Microbiome Project was launched in 2007 [18] and completed its work in 2018. The HMP was conducted in two phases over a decade to provide resources, methods, and discoveries linking human-to-microbiome interactions with health- and disease-related outcomes. The first phase of HMP (HMP1) focused on the characterization of the microbial community of numerous targeted body parts, including the oral cavity, the nose, the gut, the skin, and the vagina, from healthy adult subjects, and demonstration projects focusing on specific diseases [34,35]. The HMP1 produced abundant community resources including nucleotide sequences of microorganisms [36,37], protocols for reproducible microbiome sampling and data curation [38,39], bioinformatics methodology, and epidemiology [40,41]. One of the important findings of HMP1 was that the taxonomic composition of microbiomes alone was not strongly associated with the phenotype of the host. This tended to be better predicted by a wide range of microbial molecular functions or personalized strain-specific compositions. Based on this, the Integrative HMP (iHMP), the second phase of HMP (HMP2), was promoted to understand host–microbiome communications, including molecular mechanisms and immunity [42,43]. Unlike HMP1, which focused on the microbiome, HMP2 expanded the scope to include both the microbiome and the host in three longitudinal cohort studies on preterm birth based on vaginal microbiomes, inflammatory bowel diseases based on gut microbiomes, and prediabetes based on gut and nasal microbiomes [43]. The studies of HMP2 included both microbial and host-specific multi-omics data such as genome, epigenome, transcriptome, metabolome, and proteome. In addition to HMP, Metagenomics of the Human Intestinal Tract (MetaHIT) has produced the resources and specialty needed to understand human microbiomes.

6. Oral Microbiome

The composition of the oral microbiome differs according to its habitat [2]; the microbial population forms a unique identity and plays an important role in nutritional, defensive, and physiological activities [44]. The oral environment is a heterogeneous ecological system and it is suitable for the growth of many microorganisms due to appropriate temperature and moisture, and provides host-derived nutrients such as gingival crevicular fluid and salivary proteins [45]. The hard structures in the mouth, including tooth and dental restorations such as restoration material, dental implants, and prosthesis, provide unique, non-shedding surfaces that affect biofilm formation and calculus deposition [46,47]. In the formation of the oral microbiome, the transmission of microbes from mother to baby at birth is the initial stage, and delivery types, such as vaginally born or caesarean section, affect the composition of the microbiome [48,49]. Subsequently, feeding type also influences the composition of the oral microbiome [50], and the eruption of a tooth makes a new environment for microbial colonization [51]. After tooth eruption, the composition of the oral microbiome becomes increasingly complex with age, and transition from deciduous to permanent teeth significantly alters the dynamics of the oral microbiome [52].

Human Oral Microbiome Database (HOMD)

The main purpose of HMP was to produce 1000 microbial reference genomes, including about 300 among the oral microbes [53]. As a national resource project, the data of HMP were released rapidly, and a Data Analysis and Coordinating Center was established. Additionally, comprehensive analyses of body site-specific data from HMP1, including oral, gut, skin, and vagina data, and taxonomic classification of newly identified microbes, were needed. The HOMD is the first oral-specific public database providing the scientific community with comprehensive information on oral microbiome species [54]. The HOMD database and web-based interface were set up under the Foundation for the Oral Microbiome and Metagenome from the National Institute of Dental and Craniofacial Research, and were organized based on taxonomic classifications, which were identified based on 16S rRNA sequencing data [2,54]. The named oral species and taxa identified in 16S rRNA sequencing were placed dependent on their 16S rRNA types and their unique human oral taxon (HOT) number. The HOT interconnects phylotypes, phenotypes, and the genomic, bibliographic, and clinical information of each taxon.

7. Host–Oral Microbiome Interactions in Health

The oral cavity has evolved to improve oral health and fosters highly personalized microbiomes that exist dynamically in balance with the host. The symbiotic relationship between host and microbiome maintains microbial homeostasis; however, dysbiosis, a breakdown of the microbial homeostasis, induces oral disease and increases the risk for systemic diseases. The inseparable relationship between the host and microbiome is formed over a long time by facing various changes that force the adaptation of the oral microbiome to the new environment [55]. A bidirectional relation is characterized by the microbe providing the host with abilities it lacks alone, while the host provides an appropriate environment for microbial growth [56]. The host factors can positively affect the microbiome, making balance and diversity between the species, thus inducing symbiosis and an absence of pathology. On the contrary, the host can also create a negative influence [56,57]. This co-evolution between the host and microbiome succeeded in achieving a complex biological process in which the existence of independent entities would be impossible. The mutual benefits from the maintenance of a balanced host–oral microbiome ecology can be distorted to induce a shift from a healthy and symbiotic relation to a pathologic and dysbiotic one [58,59]. This distortion can result from changes in the oral microbiome as well as in the host [60]. Even though the host and the microbiome are equivalent factors, early studies have focused on finding the pathological oral microbiome, and the role of the host in maintaining a healthy oral microbiome was overlooked. Based on this, recent research trends have moved to focus on the host factors and combined the role of host–oral microbiome in the development of a healthy and balanced oral ecology, and extended to systemic disease and oral disease [61,62]. The host factors are largely classified into two, intrinsic and extrinsic, factors (Table 1). The oral microbiome is associated with a variety of oral diseases. Recently, there has been growing evidence that the oral microbiome is closely related to physical conditions, with many intermediate host factors [63].

Table 1. Host Factors to Modulate the Oral Microbiome.

Factor	Reference
Genetics	- Genetic polymorphism in miRNA202 is involved in hBD1 salivary level as well as caries experience [64] - Genes expressed in dental enamel development are associated with molar–incisor hypomineralization [65] - GLUT2 and TAS1R2 genotypes individually and in combination are associated with caries risk [66] - Host genetic control of the oral microbiome in health and disease [67] - Microbial abundance and some aspects of the microbial population structure are influenced by heritable traits in saliva [68]
Immunity	- Immune cell network mediating immune surveillance at oral mucosa and gingiva [69,70] - The innate host response in caries and periodontitis [71] - Secretory immunity with special reference to the oral cavity [72]
Attachment surface	- Surface properties influence oral biofilm formation [73] - Differences in relation to the microbial diversity of modified resins during the initial phase of biofilm maturation [74] - Biomaterial-associated infection of implants and devices [46]
Diet	- Vegan diet influences on the human salivary microbiota [75] - Short- and medium-chain fatty acids exhibit antimicrobial activity for oral microorganisms [76]
Cigarette smoking	- Smoking decreases structural and functional resilience in the subgingival ecosystem [77] - Firmicutes were statistically elevated in smokers at the expense of Proteobacteria and Fusobacteria in non-smokers [78] - Tobacco smoking affects the salivary gram-positive bacterial population [79]
Alcohol	- Alcohol affects to the oral microbiome composition [80–82]
Oral hygiene	- Toothbrushing frequency is related to the incidence and increment of dental caries [83]
Socioeconomic status	- Socioeconomic factors, such as education and income, are associated with disparities in the prevalence and severity of periodontal disease [84] - A strong association between cariogenic bacteria and socioeconomic status was found [85] - Differences in socioeconomic status were reflected in the bacterial profile of saliva [86]

8. Potential Clinical Application of Oral Microbiomes

With various "omics" studies, information on the composition of oral microbiomes is available. This vast amount of oral microbiome data, which were procured via HMP, could be the fundamental basis of clinical applications including early diagnosis, predictive treatment, and prevention [87,88]. The general microbial screening for diagnosis is performed using saliva and site-specific screening with gingival crevicular fluid and dental biofilm [89]. Saliva is a useful diagnostic fluid, providing the overall microbiome and proteome or metabolomic data from bacterial metabolic or host inflammatory products for personalized monitoring. This combined information from saliva can be used to predict susceptibility to oral diseases, including dental caries or periodontitis, with higher specificity [89,90]. Microbial screening of the mouth can be applied not only with oral diseases, but also with systemic diseases due to their reciprocal association.

9. Oral Disease and Systemic Disease

The commensal microbiome plays an important role in maintaining oral and systemic health. The breakdown of the microbial balance induces oral pathologic conditions such as periodontal disease, dental caries, and endodontic disease, which are associated with systemic diseases including diabetes [91], cardiovascular disease (CVD) [92], respiratory disease [93], and cancer [94] (Figure 3). The links between oral diseases and systemic health are complicated and bidirectional in many ways [95]. Among many oral diseases, periodontitis has a close relationship with non-communicable diseases (NCDs); particularly, diabetes and CVD. When periodontitis is left untreated, it could lead to the loss of periodontal supporting tissue due to microbial infection. Oral pathologic microbiomes could release virulence factors, inducing an inflammatory response, and invade the body through pathogenic lesions, which increases the risk of exacerbating NCDs [96].

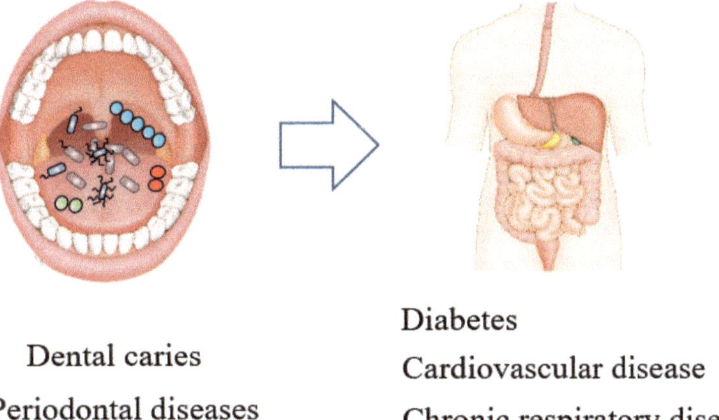

Figure 3. The oral microbiome affects both oral and systemic diseases. Adapted from Cho et al. (2021).

9.1. Non-Communicable Diseases

NCDs are diseases that are not transmissible between people, and are mainly chronic conditions with slow progression [97]. Prevalent NCDs include diabetes mellitus (DM), CVD, chronic respiratory diseases, and cancer [98]. The prevalence of NCDs is globally increasing with an aging population, which makes it a significant burden to the healthcare sector [99].

9.1.1. Cardiovascular Disease

CVD is a general term for atherosclerotic diseases, and atherosclerotic CVD is one of the leading causes of death in the world [100]. Generally, circulating leukocytes in the blood vessels do not adhere to the endothelium under normal conditions. However, when an inflammatory response occurs, various adhesion molecules, such as vascular cell adhesion molecule 1 (VCAM-1), intercellular adhesion molecule 1 (ICAM-1), and P-selectin, are expressed in the endothelial cells, inducing the adhesion of leukocytes [101]. When leukocytes adhere to the endothelium, they migrate into the intima, and diapedesis occurs via the expression of matrix metalloproteinases [102]. Several studies regarding the link between periodontitis and CVD have reported their positive association [100,105,106]. Several risk factors including smoking, diabetes, hypertension, and hypercholesterolemia have been focused on as the contributory factors in atherosclerosis; however, recent studies have suggested that immune and inflammatory mechanisms caused by oral bacteria could be an important factor in atherosclerosis [103,104]. Oral bacteria, such as periodontal and cariogenic pathogens, are known as etiological factors in the development of atherosclerosis, which is the first step of CVD. To investigate the association between oral microbiomes and atherosclerosis, several studies have been performed examining atheromatous lesions with various molecular biologic techniques. Fundamentally, periodontal pathogens were cultured in atheromatous plaque [105], and fluorescence in situ hybridization [106], DNA–DNA hybridization [107], and real-time PCR [108] results showed the presence of oral pathogens in atheromatous lesions. Based on these results, oral microbiome studies have been widely conducted on oral bacteria-induced atherosclerotic CVD (Table 2).

Table 2. Oral Microbiome Related to CVD.

Mechanism	Organism	Result	Reference
Endothelial cell invasion	P. gingivalis (P.g.) strain 381	Infection of human aortic endothelial cells with invasive P. g. strain 381 resulted in the upregulation of 68 genes that code for the pro-inflammatory cytokines, adhesion molecules, and chemokines. In addition, P. g. induces procoagulant effects including enhanced tissue factor expression and activity, and suppression of tissue factor pathway inhibitors	[109]
	F.nucleatum (F.n.)	Co-infection with F. n. resulted in a 2–20-fold increase in the invasion of endothelial cells by P. g. strains	[110]
Endothelial cell activation	A. actinomycenemcomitans (A.a.)	A.a. infection in apoliopoprotein E-deficient mice increased expressions of ICAM-1, E-selectin, P-selectin, MCP-1, chemokine (C-C motif) ligand 19 (CCL19), CCL21, and CCR7 in the aorta	[111]
	P.g.	Coculture of endothelial cells with P.g. increased ICAM-1, VCAM-1 and P-and E-selectins	[112,113]
Oxidative stress-mediated mechanism	P.g.	P.g. cleaves apoB-100 and increases the expression of apoM in LDL in whole blood	[114]
Metalloproteinase-mediated mechanism	A.a.	A.a. induces MMP-9 expression and proatherogenic lipoprotein profile in apoE-deficient mice	[115]

Table 2. *Cont.*

Mechanism	Organism	Result	Reference
Toll-like receptors-mediated mechanism	A.a.	A.a. infection of apolipoprotein E-deficient mice resulted in increased expression in the aorta of TLR2 and TLR4	[111]
	P.g.	P.g. stimulates the expression of TLR2 and TLR 4 on the surface of endothelial cells. Endothelial cells incubated with P. g. LPS expressed ICAM-1 and VCAM-1 and antibodies against TLR2	[116,117]
Acceleration of the progress of atherosclerosis	P.g.	P.g. accelerates the atherosclerosis in apolipoprotein E-null mice	[118–120]

9.1.2. Diabetes Mellitus

Periodontitis and DM are representative chronic and high-prevalence diseases in the dental and medical fields, respectively [121]. DM is a metabolic disorder characterized by prolonged high blood glucose level, which could lead to systemic complications such as CVD and circulatory problems, including peripheral vascular disease. In 2017, the International Diabetic Federation listed periodontitis as a risk factor of DM [122]. The majority of studies on DM and periodontitis have focused on type 2 DM [123,124], and the relationship of type 1 DM with periodontal disease in young patients [125]. The association between these two diseases has been studied for many years, and it is believed that the two have bidirectional links, implying that DM is a risk factor of periodontitis and periodontitis adversely affects glycemic control [126–128]. It has been known that the oral microbiota plays an important role in the relationship between DM and periodontitis because it affects blood glycemic control [129]. Disturbances in the oral microbiome are considered to be factors of periodontal disease initiation and progression and DM [130], and several studies have been conducted to understand this cause-and-effect relationship (Table 3). Matsha et al. showed the alteration in the composition of oral microbiomes across glycemic status as well as different stages of periodontal disease using 16S rDNA sequencing in dental plaque samples from South Africa [91]. Additionally, other studies with high-throughput metagenomic sequencing (16S rDNA or rRNA) of oral microbiomes also demonstrated the role of the oral microbiome in the development of DM [131–133]. Recently, Preshaw et al. reported that the treatment of periodontitis could reduce inflammation in DM patients, indicating that diabetes and periodontitis together increase systemic inflammation [134]. Similarly, a systematic review has proven that periodontal treatment, such as scaling and root planing, improves the glycemic control in DM [135]. In contrast, some reports raised the question about the effect periodontal treatment has on the glycemic control of DM patients [136,137], and showed there is no difference in oral microbiota between those with and those without DM [138,139]. These contrasting results might be due to the different types of detection methods, sampling conditions, and analysis techniques used; therefore, a comprehensive understanding with more controlled procedures and analyses would be necessary to explain the link between oral microbiome and DM.

Table 3. Oral Microbiome's Relationship with DM.

Organism	Result	Reference
P.g.	P.g. triggers periodontal tissue destruction and increases insulin resistance	[140,141]
	P.g. with type II fimbriae is a critical infectious factor in the deterioration of periodontitis with DM	[142]
	P.g. is involved with insulin resistance in DM	[143–145]

Table 3. *Cont.*

Organism	Result	Reference
Actinobacteria	Higher abundances of taxa in the phylum Actinobacteria were associated with lower diabetes risk	[131]
A.a.	A.a. was associated with periodontitis in DM patients	[146]
P.g., T. forsythia, T. denticola	Poor glycemic control is associated with increased proportions of red-complex microbes	[144]

10. Conclusions and Future Perspective

In this review, we explored the close relationship between oral microbiomes and host health. The novel technological advances in sequencing have greatly accelerated our ability to identify the microbiomes present in clinical samples taken from the oral cavity at the species level. The accumulation of knowledge about oral microbiomes is driving this research in new directions, and extended analysis of transcript (transcriptome), protein (proteome), and metabolic products (metabolome) provides insight into host–microbial interaction in oral and systemic diseases. The current state of this oral microbiome research, which has been reported so far, shows that oral diseases are complex diseases reflecting changes in microbial components and host immune responses, and are interrelated with systemic health. The combined study with multi-omics data from a host and their microbiome will facilitate advances in personalized medicine (Figure 4).

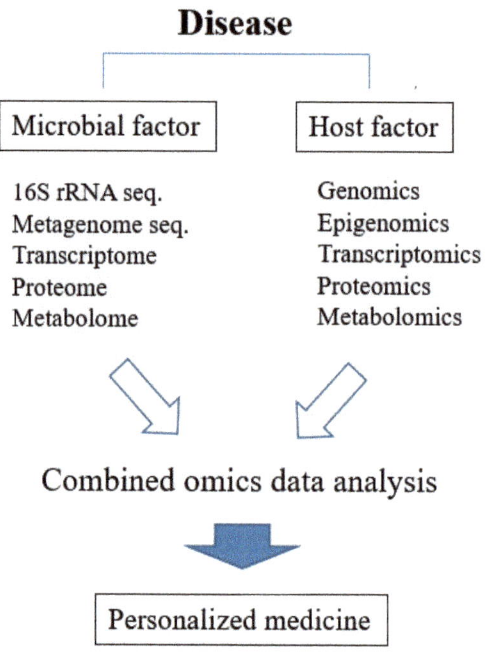

Figure 4. Personalized medicine with host–microbiome combined analysis.

Author Contributions: Conceptualization, Y.-D.C. and Y.-J.S.; methodology, Y.-D.C. and Y.-J.S.; software, Y.-D.C. and K.-H.K.; validation, Y.-D.C., K.-H.K., and Y.-J.S.; investigation, Y.-D.C. and Y.-J.S.; resources, Y.-D.C. and Y.-J.S.; original draft preparation, Y.-D.C. and Y.-J.S.; review and editing, Y.-D.C. and Y.-J.S.; supervision, Y.-M.L., Y.K., and Y.-J.S. All authors have read and agreed to the published version of the manuscript.

Funding: This work was supported by the National Research Foundation of Korea (NRF) grant funded by the Korea government (MSIT) (No. NRF-2020R1A2C1007536/2020R1C1C1005830).

Institutional Review Board Statement: Not applicable.

Informed Consent Statement: Not applicable.

Data Availability Statement: Not applicable.

Conflicts of Interest: The authors report no declarations of interest.

References

1. Priya Nimish Deo, R.D. Oral microbiome: Unveiling the fundamentals. *J. Oral Maxillofac. Pathol.* **2019**, *23*, 122–128.
2. Dewhirst, F.E.; Chen, T.; Izard, J.; Paster, B.J.; Tanner, A.C.R.; Yu, W.H.; Lakshmanan, A.; Wade, W.G. The Human Oral Microbiome. *J. Bacteriol.* **2010**, *192*, 5002–5017. [CrossRef] [PubMed]
3. Lu, M.Y.; Xuan, S.Y.; Wang, Z. Oral microbiota: A new view of body health. *Food Sci. Hum. Well.* **2019**, *8*, 8–15. [CrossRef]
4. Aas, J.A.; Paster, B.J.; Stokes, L.N.; Olsen, I.; Dewhirst, F.E. Defining the normal bacterial flora of the oral cavity. *J. Clin. Microbiol.* **2005**, *43*, 5721–5732. [CrossRef] [PubMed]
5. Lim, Y.; Totsika, M.; Morrison, M.; Punyadeera, C. Oral Microbiome: A New Biomarker Reservoir for Oral and Oropharyngeal Cancers. *Theranostics* **2017**, *7*, 4313–4321. [CrossRef] [PubMed]
6. Zhao, H.S.; Chu, M.; Huang, Z.W.; Yang, X.; Ran, S.J.; Hu, B.; Zhang, C.P.; Liang, J.P. Variations in oral microbiota associated with oral cancer. *Sci. Rep.* **2017**, *7*. [CrossRef] [PubMed]
7. Liu, S.L.; Bauer, M.E.; Marsh, P. Advancements toward a systems level understanding of the human oral microbiome. *Front. Cell. Infect. Microbiol.* **2014**, *4*. [CrossRef]
8. Ogunrinola, G.A.; Oyewale, J.O.; Oshamika, O.O.; Olasehinde, G.I. The Human Microbiome and Its Impacts on Health. *Int. J. Microbiol.* **2020**, *2020*, 8045646. [CrossRef]
9. Kilian, M.; Chapple, I.L.; Hannig, M.; Marsh, P.D.; Meuric, V.; Pedersen, A.M.; Tonetti, M.S.; Wade, W.G.; Zaura, E. The oral microbiome—An update for oral healthcare professionals. *Br. Dent. J.* **2016**, *221*, 657–666. [CrossRef] [PubMed]
10. Aguiar-Pulido, V.; Huang, W.; Suarez-Ulloa, V.; Cickovski, T.; Mathee, K.; Narasimhan, G. Metagenomics, Metatranscriptomics, and Metabolomics Approaches for Microbiome Analysis. *Evol. Bioinform. Online* **2016**, *12*, 5–16. [CrossRef]
11. Malla, M.A.; Dubey, A.; Kumar, A.; Yadav, S.; Hashem, A.; Abd Allah, E.F. Exploring the Human Microbiome: The Potential Future Role of Next-Generation Sequencing in Disease Diagnosis and Treatment. *Front. Immunol.* **2018**, *9*, 2868. [CrossRef] [PubMed]
12. Lane, N. The unseen world: Reflections on Leeuwenhoek (1677) 'Concerning little animals'. *Philos. Trans. R. Soc. Lond. B Biol. Sci.* **2015**, *370*. [CrossRef] [PubMed]
13. Rajfer, J. Antonie van Leeuwenhoek (1632–1723). *Invest. Urol.* **1976**, *14*, 83. [PubMed]
14. Hiergeist, A.; Glasner, J.; Reischl, U.; Gessner, A. Analyses of Intestinal Microbiota: Culture versus Sequencing. *ILAR J.* **2015**, *56*, 228–240. [CrossRef] [PubMed]
15. Ha, C.W.Y.; Devkota, S. The new microbiology: Cultivating the future of microbiome-directed medicine. *Am. J. Physiol. Gastrointest. Liver Physiol.* **2020**, *319*, G639–G645. [CrossRef] [PubMed]
16. Lander, E.S.; Linton, L.M.; Birren, B.; Nusbaum, C.; Zody, M.C.; Baldwin, J.; Devon, K.; Dewar, K.; Doyle, M.; FitzHugh, W.; et al. Initial sequencing and analysis of the human genome. *Nature* **2001**, *409*, 860–921. [CrossRef] [PubMed]
17. Carrasco-Ramiro, F.; Peiro-Pastor, R.; Aguado, B. Human genomics projects and precision medicine. *Gene Ther.* **2017**, *24*, 551–561. [CrossRef] [PubMed]
18. Turnbaugh, P.J.; Ley, R.E.; Hamady, M.; Fraser-Liggett, C.M.; Knight, R.; Gordon, J.I. The human microbiome project. *Nature* **2007**, *449*, 804–810. [CrossRef]
19. Zaura, E. Next-generation sequencing approaches to understanding the oral microbiome. *Adv. Dent. Res.* **2012**, *24*, 81–85. [CrossRef] [PubMed]
20. Werner, J.J.; Zhou, D.; Caporaso, J.G.; Knight, R.; Angenent, L.T. Comparison of Illumina paired-end and single-direction sequencing for microbial 16S rRNA gene amplicon surveys. *ISME J.* **2012**, *6*, 1273–1276. [CrossRef] [PubMed]
21. Hugenholtz, P.; Goebel, B.M.; Pace, N.R. Impact of culture-independent studies on the emerging phylogenetic view of bacterial diversity. *J. Bacteriol.* **1998**, *180*, 4765–4774. [CrossRef] [PubMed]
22. Hebert, P.D.; Cywinska, A.; Ball, S.L.; deWaard, J.R. Biological identifications through DNA barcodes. *Proc. Biol. Sci.* **2003**, *270*, 313–321. [CrossRef] [PubMed]
23. Ruppert, K.M.; Kline, R.J.; Rahman, M.S. Past, present, and future perspectives of environmental DNA (eDNA) metabarcoding: A systematic review in methods, monitoring, and applications of global eDNA. *Glob. Ecol. Conserv.* **2019**, *17*. [CrossRef]
24. Simon, C.; Daniel, R. Metagenomic analyses: Past and future trends. *Appl. Environ. Microbiol.* **2011**, *77*, 1153–1161. [CrossRef]
25. Hettich, R.L.; Pan, C.; Chourey, K.; Giannone, R.J. Metaproteomics: Harnessing the power of high performance mass spectrometry to identify the suite of proteins that control metabolic activities in microbial communities. *Anal. Chem.* **2013**, *85*, 4203–4214. [CrossRef]
26. Bostanci, N.; Bao, K. Contribution of proteomics to our understanding of periodontal inflammation. *Proteomics* **2017**, *17*. [CrossRef]

27. Bostanci, N.; Grant, M.; Bao, K.; Silbereisen, A.; Hetrodt, F.; Manoil, D.; Belibasakis, G.N. Metaproteome and metabolome of oral microbial communities. *Periodontol. 2000* **2021**, *85*, 46–81. [CrossRef] [PubMed]

28. Risely, A. Applying the core microbiome to understand host-microbe systems. *J. Anim. Ecol.* **2020**, *89*, 1549–1558. [CrossRef]

29. Hamady, M.; Knight, R. Microbial community profiling for human microbiome projects: Tools, techniques, and challenges. *Genome Res.* **2009**, *19*, 1141–1152. [CrossRef]

30. Zaura, E.; Keijser, B.J.; Huse, S.M.; Crielaard, W. Defining the healthy "core microbiome" of oral microbial communities. *BMC Microbiol.* **2009**, *9*, 259. [CrossRef] [PubMed]

31. Chow, J.; Lee, S.M.; Shen, Y.; Khosravi, A.; Mazmanian, S.K. Host-bacterial symbiosis in health and disease. *Adv. Immunol.* **2010**, *107*, 243–274. [CrossRef] [PubMed]

32. Keijser, B.J.; Zaura, E.; Huse, S.M.; van der Vossen, J.M.; Schuren, F.H.; Montijn, R.C.; ten Cate, J.M.; Crielaard, W. Pyrosequencing analysis of the oral microflora of healthy adults. *J. Dent. Res.* **2008**, *87*, 1016–1020. [CrossRef] [PubMed]

33. Bao, K.; Li, X.; Poveda, L.; Qi, W.; Selevsek, N.; Gumus, P.; Emingil, G.; Grossmann, J.; Diaz, P.I.; Hajishengallis, G.; et al. Proteome and Microbiome Mapping of Human Gingival Tissue in Health and Disease. *Front. Cell. Infect. Microbiol.* **2020**, *10*, 588155. [CrossRef] [PubMed]

34. Human Microbiome Project Consortium. Structure, function and diversity of the healthy human microbiome. *Nature* **2012**, *486*, 207–214. [CrossRef] [PubMed]

35. Human Microbiome Project Consortium. A framework for human microbiome research. *Nature* **2012**, *486*, 215–221. [CrossRef] [PubMed]

36. Human Microbiome Jumpstart Reference Strains Consortium; Nelson, K.E.; Weinstock, G.M.; Highlander, S.K.; Worley, K.C.; Creasy, H.H.; Wortman, J.R.; Rusch, D.B.; Mitreva, M.; Sodergren, E.; et al. A catalog of reference genomes from the human microbiome. *Science* **2010**, *328*, 994–999. [CrossRef]

37. Wylie, K.M.; Truty, R.M.; Sharpton, T.J.; Mihindukulasuriya, K.A.; Zhou, Y.; Gao, H.; Sodergren, E.; Weinstock, G.M.; Pollard, K.S. Novel bacterial taxa in the human microbiome. *PLoS ONE* **2012**, *7*, e35294. [CrossRef] [PubMed]

38. Aagaard, K.; Petrosino, J.; Keitel, W.; Watson, M.; Katancik, J.; Garcia, N.; Patel, S.; Cutting, M.; Madden, T.; Hamilton, H.; et al. The Human Microbiome Project strategy for comprehensive sampling of the human microbiome and why it matters. *FASEB J.* **2013**, *27*, 1012–1022. [CrossRef] [PubMed]

39. Ward, D.V.; Gevers, D.; Giannoukos, G.; Earl, A.M.; Methe, B.A.; Sodergren, E.; Feldgarden, M.; Ciulla, D.M.; Tabbaa, D.; Arze, C.; et al. Evaluation of 16S rDNA-Based Community Profiling for Human Microbiome Research. *PLoS ONE* **2012**, *7*. [CrossRef]

40. Gevers, D.; Pop, M.; Schloss, P.D.; Huttenhower, C. Bioinformatics for the Human Microbiome Project. *PLoS Comp. Biol.* **2012**, *8*. [CrossRef] [PubMed]

41. Markowitz, V.M.; Chen, I.M.A.; Chu, K.; Szeto, E.; Palaniappan, K.; Jacob, B.; Ratner, A.; Liolios, K.; Pagani, I.; Huntemann, M.; et al. IMG/M-HMP: A Metagenome Comparative Analysis System for the Human Microbiome Project. *PLoS ONE* **2012**, *7*. [CrossRef]

42. Integrative, H.M.P. The Integrative Human Microbiome Project: Dynamic Analysis of Microbiome-Host Omics Profiles during Periods of Human Health and Disease. *Cell Host Microbe* **2014**, *16*, 276–289. [CrossRef]

43. Proctor, L.M.; Creasy, H.H.; Fettweis, J.M.; Lloyd-Price, J.; Mahurkar, A.; Zhou, W.Y.; Buck, G.A.; Snyder, M.P.; Strauss, J.F.; Weinstock, G.M.; et al. The Integrative Human Microbiome Project. *Nature* **2019**, *569*, 641–648. [CrossRef]

44. Franzosa, E.A.; Huang, K.; Meadow, J.F.; Gevers, D.; Lemon, K.P.; Bohannan, B.J.M.; Huttenhower, C. Identifying personal microbiomes using metagenomic codes. *Proc. Natl. Acad. Sci. USA* **2015**, *112*, E2930–E2938. [CrossRef] [PubMed]

45. Van't Hof, W.; Veerman, E.C.; Nieuw Amerongen, A.V.; Ligtenberg, A.J. Antimicrobial defense systems in saliva. *Monogr. Oral Sci.* **2014**, *24*, 40–51. [CrossRef]

46. Busscher, H.J.; Rinastiti, M.; Siswomihardjo, W.; van der Mei, H.C. Biofilm formation on dental restorative and implant materials. *J. Dent. Res.* **2010**, *89*, 657–665. [CrossRef] [PubMed]

47. Oilo, M.; Bakken, V. Biofilm and Dental Biomaterials. *Materials* **2015**, *8*, 2887–2900. [CrossRef]

48. Dominguez-Bello, M.G.; Costello, E.K.; Contreras, M.; Magris, M.; Hidalgo, G.; Fierer, N.; Knight, R. Delivery mode shapes the acquisition and structure of the initial microbiota across multiple body habitats in newborns. *Proc. Natl. Acad. Sci. USA* **2010**, *107*, 11971–11975. [CrossRef] [PubMed]

49. Holgerson, P.L.; Harnevik, L.; Hernell, O.; Tanner, A.C.R.; Johansson, I. Mode of Birth Delivery Affects Oral Microbiota in Infants. *J. Dent. Res.* **2011**, *90*, 1183–1188. [CrossRef] [PubMed]

50. Holgerson, P.L.; Vestman, N.R.; Claesson, R.; Ohman, C.; Domellof, M.; Tanner, A.C.R.; Hernell, O.; Johansson, I. Oral Microbial Profile Discriminates Breast-fed From Formula-fed Infants. *J. Pediatr. Gastroenterol. Nutr.* **2013**, *56*, 127–136. [CrossRef]

51. Sampaio-Maia, B.; Monteiro-Silva, F. Acquisition and maturation of oral microbiome throughout childhood: An update. *Dent. Res. J. (Isfahan)* **2014**, *11*, 291–301.

52. Xu, X.; He, J.Z.; Xue, J.; Wang, Y.; Li, K.; Zhang, K.K.; Guo, Q.; Liu, X.H.; Zhou, Y.; Cheng, L.; et al. Oral cavity contains distinct niches with dynamic microbial communities. *Environ. Microbiol.* **2015**, *17*, 699–710. [CrossRef]

53. Group, N.H.W.; Peterson, J.; Garges, S.; Giovanni, M.; McInnes, P.; Wang, L.; Schloss, J.A.; Bonazzi, V.; McEwen, J.E.; Wetterstrand, K.A.; et al. The NIH Human Microbiome Project. *Genome Res.* **2009**, *19*, 2317–2323. [CrossRef] [PubMed]

54. Chen, T.; Yu, W.H.; Izard, J.; Baranova, O.V.; Lakshmanan, A.; Dewhirst, F.E. The Human Oral Microbiome Database: A web accessible resource for investigating oral microbe taxonomic and genomic information. *Database (Oxford)* **2010**, *2010*, baq013. [CrossRef] [PubMed]

55. Cross, K.L.; Chirania, P.; Xiong, W.L.; Beall, C.J.; Elkins, J.G.; Giannone, R.J.; Griffen, A.L.; Guss, A.M.; Hettich, R.L.; Joshi, S.S.; et al. Insights into the Evolution of Host Association through the Isolation and Characterization of a Novel Human Periodontal Pathobiont, Desulfobulbus oralis. *mBio* **2018**, *9*. [CrossRef] [PubMed]

56. Ulloa, P.C.; van der Veen, M.H.; Krom, B.P. Review: Modulation of the oral microbiome by the host to promote ecological balance. *Odontology* **2019**, *107*, 437–448. [CrossRef] [PubMed]

57. Zheng, D.P.; Liwinski, T.; Elinav, E. Interaction between microbiota and immunity in health and disease. *Cell Res.* **2020**, *30*, 492–506. [CrossRef] [PubMed]

58. Sanz, M.; Beighton, D.; Curtis, M.A.; Cury, J.A.; Dige, I.; Dommisch, H.; Ellwood, R.; Giacaman, R.A.; Herrera, D.; Herzberg, M.C.; et al. Role of microbial biofilms in the maintenance of oral health and in the development of dental caries and periodontal diseases. Consensus report of group 1 of the Joint EFP/ORCA workshop on the boundaries between caries and periodontal disease. *J. Clin. Periodontol.* **2017**, *44* (Suppl. S18), S5–S11. [CrossRef] [PubMed]

59. Sampaio-Maia, B.; Caldas, I.M.; Pereira, M.L.; Perez-Mongiovi, D.; Araujo, R. The Oral Microbiome in Health and Its Implication in Oral and Systemic Diseases. *Adv. Appl. Microbiol.* **2016**, *97*, 171–210. [CrossRef] [PubMed]

60. Marsh, P.D.; Devine, D.A. How is the development of dental biofilms influenced by the host? *J. Clin. Periodontol.* **2011**, *38* (Suppl. S11), 28–35. [CrossRef]

61. Chimenos-Kustner, E.; Giovannoni, M.L.; Schemel-Suarez, M. Dysbiosis as a determinant factor of systemic and oral pathology: Importance of microbiome. *Med. Clin. (Barc.)* **2017**, *149*, 305–309. [CrossRef] [PubMed]

62. Zhang, Y.; Wang, X.; Li, H.; Ni, C.; Du, Z.; Yan, F. Human oral microbiota and its modulation for oral health. *Biomed. Pharmacother.* **2018**, *99*, 883–893. [CrossRef] [PubMed]

63. Genco, R.J.; Borgnakke, W.S. Risk factors for periodontal disease. *Periodontol. 2000* **2013**, *62*, 59–94. [CrossRef] [PubMed]

64. Lips, A.; Antunes, L.S.; Antunes, L.A.; Abreu, J.G.B.; Barreiros, D.; Oliveira, D.S.B.; Batista, A.C.; Nelson-Filho, P.; Silva, L.; Silva, R.; et al. Genetic Polymorphisms in DEFB1 and miRNA202 Are Involved in Salivary Human beta-Defensin 1 Levels and Caries Experience in Children. *Caries Res.* **2017**, *51*, 209–215. [CrossRef] [PubMed]

65. Jeremias, F.; Koruyucu, M.; Kuchler, E.C.; Bayram, M.; Tuna, E.B.; Deeley, K.; Pierri, R.A.; Souza, J.F.; Fragelli, C.M.; Paschoal, M.A.; et al. Genes expressed in dental enamel development are associated with molar-incisor hypomineralization. *Arch. Oral Biol.* **2013**, *58*, 1434–1442. [CrossRef]

66. Kulkarni, G.V.; Chng, T.; Eny, K.M.; Nielsen, D.; Wessman, C.; El-Sohemy, A. Association of GLUT2 and TAS1R2 genotypes with risk for dental caries. *Caries Res.* **2013**, *47*, 219–225. [CrossRef] [PubMed]

67. Gomez, A.; Espinoza, J.L.; Harkins, D.M.; Leong, P.; Saffery, R.; Bockmann, M.; Torralba, M.; Kuelbs, C.; Kodukula, R.; Inman, J.; et al. Host Genetic Control of the Oral Microbiome in Health and Disease. *Cell Host Microbe* **2017**, *22*, 269–278. [CrossRef]

68. Demmitt, B.A.; Corley, R.P.; Huibregtse, B.M.; Keller, M.C.; Hewitt, J.K.; McQueen, M.B.; Knight, R.; McDermott, I.; Krauter, K.S. Genetic influences on the human oral microbiome. *BMC Genomics* **2017**, *18*, 659. [CrossRef]

69. Moutsopoulos, N.M.; Konkel, J.E. Tissue-Specific Immunity at the Oral Mucosal Barrier. *Trends Immunol.* **2018**, *39*, 276–287. [CrossRef]

70. Dutzan, N.; Konkel, J.E.; Greenwell-Wild, T.; Moutsopoulos, N.M. Characterization of the human immune cell network at the gingival barrier. *Mucosal Immunol.* **2016**, *9*, 1163–1172. [CrossRef]

71. Meyle, J.; Dommisch, H.; Groeger, S.; Giacaman, R.A.; Costalonga, M.; Herzberg, M. The innate host response in caries and periodontitis. *J. Clin. Periodontol.* **2017**, *44*, 1215–1225. [CrossRef] [PubMed]

72. Brandtzaeg, P. Secretory immunity with special reference to the oral cavity. *J. Oral Microbiol.* **2013**, *5*. [CrossRef] [PubMed]

73. Song, F.; Koo, H.; Ren, D. Effects of Material Properties on Bacterial Adhesion and Biofilm Formation. *J. Dent. Res.* **2015**, *94*, 1027–1034. [CrossRef] [PubMed]

74. De Castro, D.T.; do Nascimento, C.; Alves, O.L.; Santos, E.D.; Agnelli, J.A.M.; dos Reis, A.C. Analysis of the oral microbiome on the surface of modified dental polymers. *Arch. Oral Biol.* **2018**, *93*, 107–114. [CrossRef] [PubMed]

75. Hansen, T.H.; Kern, T.; Bak, E.G.; Kashani, A.; Allin, K.H.; Nielsen, T.; Hansen, T.; Pedersen, O. Impact of a vegan diet on the human salivary microbiota. *Sci. Rep.* **2018**, *8*. [CrossRef]

76. Huang, C.B.; Alimova, Y.; Myers, T.M.; Ebersole, J.L. Short- and medium-chain fatty acids exhibit antimicrobial activity for oral microorganisms. *Arch. Oral Biol.* **2011**, *56*, 650–654. [CrossRef] [PubMed]

77. Joshi, V.; Matthews, C.; Aspiras, M.; de Jager, M.; Ward, M.; Kumar, P. Smoking decreases structural and functional resilience in the subgingival ecosystem. *J. Clin. Periodontol.* **2014**, *41*, 1037–1047. [CrossRef] [PubMed]

78. Al-Zyoud, W.; Hajjo, R.; Abu-Siniyeh, A.; Hajjaj, S. Salivary Microbiome and Cigarette Smoking: A First of Its Kind Investigation in Jordan. *Int. J. Environ. Res. Public Health* **2020**, *17*, 256. [CrossRef]

79. Grine, G.; Royer, A.; Terrer, E.; Diallo, O.O.; Drancourt, M.; Aboudharam, G. Tobacco Smoking Affects the Salivary Gram-Positive Bacterial Population. *Front. Public Health* **2019**, *7*, 196. [CrossRef]

80. O'Grady, I.; Anderson, A.; O'Sullivan, J. The interplay of the oral microbiome and alcohol consumption in oral squamous cell carcinomas. *Oral Oncol.* **2020**, *110*. [CrossRef] [PubMed]

81. Snider, J. Alcohol Consumption, Particularly among Heavy Drinkers, May Affect Oral Microbiome Composition, Researchers Find. *J. Am. Dent. Assoc.* **2018**, *149*, 574–575.
82. Fan, X.Z.; Peters, B.A.; Jacobs, E.J.; Gapstur, S.M.; Purdue, M.P.; Freedman, N.D.; Alekseyenko, A.V.; Wu, J.; Yang, L.Y.; Pei, Z.H.; et al. Drinking alcohol is associated with variation in the human oral microbiome in a large study of American adults. *Microbiome* **2018**, *6*. [CrossRef] [PubMed]
83. Kumar, S.; Tadakamadla, J.; Johnson, N.W. Effect of Toothbrushing Frequency on Incidence and Increment of Dental Caries: A Systematic Review and Meta-Analysis. *J. Dent. Res.* **2016**, *95*, 1230–1236. [CrossRef] [PubMed]
84. Borrell, L.N.; Crawford, N.D. Socioeconomic position indicators and periodontitis: Examining the evidence. *Periodontol. 2000* **2012**, *58*, 69–83. [CrossRef] [PubMed]
85. Boyce, W.T.; Den Besten, P.K.; Stamperdahl, J.; Zhan, L.; Jiang, Y.; Adler, N.E.; Featherstone, J.D. Social inequalities in childhood dental caries: The convergent roles of stress, bacteria and disadvantage. *Soc. Sci. Med.* **2010**, *71*, 1644–1652. [CrossRef] [PubMed]
86. Belstrom, D.; Holmstrup, P.; Nielsen, C.H.; Kirkby, N.; Twetman, S.; Heitmann, B.L.; Klepac-Ceraj, V.; Paster, B.J.; Fiehn, N.E. Bacterial profiles of saliva in relation to diet, lifestyle factors, and socioeconomic status. *J. Oral Microbiol.* **2014**, *6*. [CrossRef] [PubMed]
87. Cantore, S.; Inchingolo, A.D.; Xhajanka, E.; Altini, V.; Bordea, I.R.; Dipalma, G.; Inchingolo, F. Management of patients suffering from xerostomia with a combined mouthrinse containing sea salt, xylitol and lysozyme. *J. Biol. Regul. Homeost. Agents* **2020**, *34*, 1607–1611. [CrossRef] [PubMed]
88. Chiniforush, N.; Pourhajibagher, M.; Parker, S.; Benedicenti, S.; Bahador, A.; Salagean, T.; Bordea, I.R. The Effect of Antimicrobial Photodynamic Therapy Using Chlorophyllin-Phycocyanin Mixture onEnterococcus faecalis: The Influence of Different Light Sources. *Appl. Sci.* **2020**, *10*, 4290. [CrossRef]
89. Belibasakis, G.N.; Bostanci, N.; Marsh, P.D.; Zaura, E. Applications of the oral microbiome in personalized dentistry. *Arch. Oral Biol.* **2019**, *104*, 7–12. [CrossRef]
90. Kaman, W.E.; Galassi, F.; de Soet, J.J.; Bizzarro, S.; Loos, B.G.; Veerman, E.C.; van Belkum, A.; Hays, J.P.; Bikker, F.J. Highly specific protease-based approach for detection of porphyromonas gingivalis in diagnosis of periodontitis. *J. Clin. Microbiol.* **2012**, *50*, 104–112. [CrossRef]
91. Matsha, T.E.; Prince, Y.; Davids, S.; Chikte, U.; Erasmus, R.T.; Kengne, A.P.; Davison, G.M. Oral Microbiome Signatures in Diabetes Mellitus and Periodontal Disease. *J. Dent. Res.* **2020**, *99*, 658–665. [CrossRef] [PubMed]
92. Beck, J.D.; Offenbacher, S. Systemic effects of periodontitis: Epidemiology of periodontal disease and cardiovascular disease. *J. Periodontol.* **2005**, *76*, 2089–2100. [CrossRef] [PubMed]
93. Gomes-Filho, I.S.; Passos, J.S.; Seixas da Cruz, S. Respiratory disease and the role of oral bacteria. *J. Oral Microbiol.* **2010**, *2*. [CrossRef] [PubMed]
94. Teles, F.R.F.; Alawi, F.; Castilho, R.M.; Wang, Y. Association or Causation? Exploring the Oral Microbiome and Cancer Links. *J. Dent. Res.* **2020**, *99*, 1411–1424. [CrossRef] [PubMed]
95. Dorfer, C.; Benz, C.; Aida, J.; Campard, G. The relationship of oral health with general health and NCDs: A brief review. *Int. Dent. J.* **2017**, *67* (Suppl. S2), 14–18. [CrossRef] [PubMed]
96. Scannapieco, F.A.; Cantos, A. Oral inflammation and infection, and chronic medical diseases: Implications for the elderly. *Periodontol. 2000* **2016**, *72*, 153–175. [CrossRef] [PubMed]
97. Cho, Y.D.; Kim, W.J.; Ryoo, H.M.; Kim, H.G.; Kim, K.H.; Ku, Y.; Seol, Y.J. Current advances of epigenetics in periodontology from ENCODE project: A review and future perspectives. *Clin. Epigenetics* **2021**, *13*, 92. [CrossRef] [PubMed]
98. Sanz, M.; Marco Del Castillo, A.; Jepsen, S.; Gonzalez-Juanatey, J.R.; D'Aiuto, F.; Bouchard, P.; Chapple, I.; Dietrich, T.; Gotsman, I.; Graziani, F.; et al. Periodontitis and cardiovascular diseases: Consensus report. *J. Clin. Periodontol.* **2020**, *47*, 268–288. [CrossRef]
99. Jan, S.; Laba, T.L.; Essue, B.M.; Gheorghe, A.; Muhunthan, J.; Engelgau, M.; Mahal, A.; Griffiths, U.; McIntyre, D.; Meng, Q.; et al. Action to address the household economic burden of non-communicable diseases. *Lancet* **2018**, *391*, 2047–2058. [CrossRef]
100. Barquera, S.; Pedroza-Tobias, A.; Medina, C.; Hernandez-Barrera, L.; Bibbins-Domingo, K.; Lozano, R.; Moran, A.E. Global Overview of the Epidemiology of Atherosclerotic Cardiovascular Disease. *Arch. Med. Res.* **2015**, *46*, 328–338. [CrossRef]
101. Libby, P.; Theroux, P. Pathophysiology of coronary artery disease. *Circulation* **2005**, *111*, 3481–3488. [CrossRef] [PubMed]
102. Agewall, S. Matrix metalloproteinases and cardiovascular disease. *Eur. Heart J.* **2006**, *27*, 121–122. [CrossRef] [PubMed]
103. Vita, J.A.; Loscalzo, J. Shouldering the risk factor burden: Infection, atherosclerosis, and the vascular endothelium. *Circulation* **2002**, *106*, 164–166. [CrossRef] [PubMed]
104. Spahr, A.; Klein, E.; Khuseyinova, N.; Boeckh, C.; Muche, R.; Kunze, M.; Rothenbacher, D.; Pezeshki, G.; Hoffmeister, A.; Koenig, W. Periodontal infections and coronary heart disease: Role of periodontal bacteria and importance of total pathogen burden in the Coronary Event and Periodontal Disease (CORODONT) study. *Arch. Intern. Med.* **2006**, *166*, 554–559. [CrossRef] [PubMed]
105. Kozarov, E.V.; Dorn, B.R.; Shelburne, C.E.; Dunn, W.A., Jr.; Progulske-Fox, A. Human atherosclerotic plaque contains viable invasive Actinobacillus actinomycetemcomitans and Porphyromonas gingivalis. *Arterioscler. Thromb. Vasc. Biol.* **2005**, *25*, e17–e18. [CrossRef] [PubMed]
106. Cavrini, F.; Sambri, V.; Moter, A.; Servidio, D.; Marangoni, A.; Montebugnoli, L.; Foschi, F.; Prati, C.; Di Bartolomeo, R.; Cevenini, R. Molecular detection of Treponema denticola and Porphyromonas gingivalis in carotid and aortic atheromatous plaques by FISH: Report of two cases. *J. Med. Microbiol.* **2005**, *54*, 93–96. [CrossRef] [PubMed]

107. Elkaim, R.; Dahan, M.; Kocgozlu, L.; Werner, S.; Kanter, D.; Kretz, J.G.; Tenenbaum, H. Prevalence of periodontal pathogens in subgingival lesions, atherosclerotic plaques and healthy blood vessels: A preliminary study. *J. Periodontal Res.* **2008**, *43*, 224–231. [CrossRef]

108. Gaetti-Jardim, E.; Marcelino, S.L.; Feitosa, A.C.R.; Romito, G.A.; Avila-Campos, M.J. Quantitative detection of periodontopathic bacteria in atherosclerotic plaques from coronary arteries. *J. Med. Microbiol.* **2009**, *58*, 1568–1575. [CrossRef] [PubMed]

109. Chou, H.H.; Yumoto, H.; Davey, M.; Takahashi, Y.; Miyamoto, T.; Gibson, F.C., 3rd; Genco, C.A. Porphyromonas gingivalis fimbria-dependent activation of inflammatory genes in human aortic endothelial cells. *Infect. Immun.* **2005**, *73*, 5367–5378. [CrossRef] [PubMed]

110. Saito, Y.; Fujii, R.; Nakagawa, K.I.; Kuramitsu, H.K.; Okuda, K.; Ishihara, K. Stimulation of Fusobacterium nucleatum biofilm formation by Porphyromonas gingivalis. *Oral Microbiol. Immunol.* **2008**, *23*, 1–6. [CrossRef] [PubMed]

111. Zhang, T.; Kurita-Ochiai, T.; Hashizume, T.; Du, Y.; Oguchi, S.; Yamamoto, M. Aggregatibacter actinomycetemcomitans accelerates atherosclerosis with an increase in atherogenic factors in spontaneously hyperlipidemic mice. *FEMS Immunol. Med. Microbiol.* **2010**, *59*, 143–151. [CrossRef] [PubMed]

112. Khlgatian, M.; Nassar, H.; Chou, H.H.; Gibson, F.C., 3rd; Genco, C.A. Fimbria-dependent activation of cell adhesion molecule expression in Porphyromonas gingivalis-infected endothelial cells. *Infect. Immun.* **2002**, *70*, 257–267. [CrossRef] [PubMed]

113. Takahashi, Y.; Davey, M.; Yumoto, H.; Gibson, F.C., 3rd; Genco, C.A. Fimbria-dependent activation of pro-inflammatory molecules in Porphyromonas gingivalis infected human aortic endothelial cells. *Cell. Microbiol.* **2006**, *8*, 738–757. [CrossRef] [PubMed]

114. Bengtsson, T.; Karlsson, H.; Gunnarsson, P.; Skoglund, C.; Elison, C.; Leanderson, P.; Lindahl, M. The periodontal pathogen Porphyromonas gingivalis cleaves apoB-100 and increases the expression of apoM in LDL in whole blood leading to cell proliferation. *J. Intern. Med.* **2008**, *263*, 558–571. [CrossRef] [PubMed]

115. Tuomainen, A.M.; Jauhiainen, M.; Kovanen, P.T.; Metso, J.; Paju, S.; Pussinen, P.J. Aggregatibacter actinomycetemcomitans induces MMP-9 expression and proatherogenic lipoprotein profile in apoE-deficient mice. *Microb. Pathog.* **2008**, *44*, 111–117. [CrossRef] [PubMed]

116. Yumoto, H.; Chou, H.H.; Takahashi, Y.; Davey, M.; Gibson, F.C., 3rd; Genco, C.A. Sensitization of human aortic endothelial cells to lipopolysaccharide via regulation of Toll-like receptor 4 by bacterial fimbria-dependent invasion. *Infect. Immun.* **2005**, *73*, 8050–8059. [CrossRef]

117. Nakamura, N.; Yoshida, M.; Umeda, M.; Huang, Y.; Kitajima, S.; Inoue, Y.; Ishikawa, I.; Iwai, T. Extended exposure of lipopolysaccharide fraction from Porphyromonas gingivalis facilitates mononuclear cell adhesion to vascular endothelium via Toll-like receptor-2 dependent mechanism. *Atherosclerosis* **2008**, *196*, 59–67. [CrossRef] [PubMed]

118. Gibson, F.C., 3rd; Hong, C.; Chou, H.H.; Yumoto, H.; Chen, J.; Lien, E.; Wong, J.; Genco, C.A. Innate immune recognition of invasive bacteria accelerates atherosclerosis in apolipoprotein E-deficient mice. *Circulation* **2004**, *109*, 2801–2806. [CrossRef]

119. Brodala, N.; Merricks, E.P.; Bellinger, D.A.; Damrongsri, D.; Offenbacher, S.; Beck, J.; Madianos, P.; Sotres, D.; Chang, Y.L.; Koch, G.; et al. Porphyromonas gingivalis bacteremia induces coronary and aortic atherosclerosis in normocholesterolemic and hypercholesterolemic pigs. *Arterioscler. Thromb. Vasc. Biol.* **2005**, *25*, 1446–1451. [CrossRef]

120. Lalla, E.; Lamster, I.B.; Hofmann, M.A.; Bucciarelli, L.; Jerud, A.P.; Tucker, S.; Lu, Y.; Papapanou, P.N.; Schmidt, A.M. Oral infection with a periodontal pathogen accelerates early atherosclerosis in apolipoprotein E-null mice. *Arterioscler. Thromb. Vasc. Biol.* **2003**, *23*, 1405–1411. [CrossRef]

121. Pizzo, G.; Guiglia, R.; Lo Russo, L.; Campisi, G. Dentistry and internal medicine: From the focal infection theory to the periodontal medicine concept. *Eur. J. Intern. Med.* **2010**, *21*, 496–502. [CrossRef] [PubMed]

122. Sanz, M.; Ceriello, A.; Buysschaert, M.; Chapple, I.; Demmer, R.T.; Graziani, F.; Herrera, D.; Jepsen, S.; Lione, L.; Madianos, P.; et al. Scientific evidence on the links between periodontal diseases and diabetes: Consensus report and guidelines of the joint workshop on periodontal diseases and diabetes by the International Diabetes Federation and the European Federation of Periodontology. *J. Clin. Periodontol.* **2018**, *45*, 138–149. [CrossRef] [PubMed]

123. Padhi, S.; Nayak, A.K.; Behera, A. Type II diabetes mellitus: A review on recent drug based therapeutics. *Biomed. Pharmacother.* **2020**, *131*, 110708. [CrossRef]

124. Chatterjee, S.; Khunti, K.; Davies, M.J. Type 2 diabetes. *Lancet* **2017**, *389*, 2239–2251. [CrossRef]

125. Lalla, E.; Cheng, B.; Lal, S.; Kaplan, S.; Softness, B.; Greenberg, E.; Goland, R.S.; Lamster, I.B. Diabetes mellitus promotes periodontal destruction in children. *J. Clin. Periodontol.* **2007**, *34*, 294–298. [CrossRef] [PubMed]

126. Taylor, G.W. Bidirectional interrelationships between diabetes and periodontal diseases: An epidemiologic perspective. *Ann. Periodontol.* **2001**, *6*, 99–112. [CrossRef] [PubMed]

127. Negrato, C.A.; Tarzia, O.; Jovanovic, L.; Chinellato, L.E.M. Periodontal disease and diabetes mellitus. *J. Oral Microbiol.* **2013**, *21*, 1–12. [CrossRef] [PubMed]

128. Taylor, J.J.; Preshaw, P.M.; Lalla, E. A review of the evidence for pathogenic mechanisms that may link periodontitis and diabetes. *J. Clin. Periodontol.* **2013**, *40* (Suppl. S14), S113–S134. [CrossRef]

129. Mealey, B.L.; Oates, T.W.; American Academy of, P. Diabetes mellitus and periodontal diseases. *J. Periodontol.* **2006**, *77*, 1289–1303. [CrossRef]

130. Signorini, L.; Inchingolo, A.D.; Santacroce, L.; Xhajanka, E.; Altini, V.; Bordea, I.R.; Dipalma, G.; Cantore, S.; Inchingolo, F. Efficacy of combined sea salt based oral rinse with xylitol in improving healing process and oral hygiene among diabetic population after oral surgery. *J. Biol. Regul. Homeost. Agents* **2020**, *34*, 1617–1622. [CrossRef] [PubMed]

131. Long, J.; Cai, Q.; Steinwandel, M.; Hargreaves, M.K.; Bordenstein, S.R.; Blot, W.J.; Zheng, W.; Shu, X.O. Association of oral microbiome with type 2 diabetes risk. *J. Periodontal Res.* **2017**, *52*, 636–643. [CrossRef] [PubMed]
132. Ogawa, T.; Honda-Ogawa, M.; Ikebe, K.; Notomi, Y.; Iwamoto, Y.; Shirobayashi, I.; Hata, S.; Kibi, M.; Masayasu, S.; Sasaki, S.; et al. Characterizations of oral microbiota in elderly nursing home residents with diabetes. *J. Oral Sci.* **2017**, *59*, 549–555. [CrossRef] [PubMed]
133. Saeb, A.T.M.; Al-Rubeaan, K.A.; Aldosary, K.; Raja, G.K.U.; Mani, B.; Abouelhoda, M.; Tayeb, H.T. Relative reduction of biological and phylogenetic diversity of the oral microbiota of diabetes and pre-diabetes patients. *Microb. Pathog.* **2019**, *128*, 215–229. [CrossRef] [PubMed]
134. Preshaw, P.M.; Taylor, J.J.; Jaedicke, K.M.; De Jager, M.; Bikker, J.W.; Selten, W.; Bissett, S.M.; Whall, K.M.; van de Merwe, R.; Areibi, A.; et al. Treatment of periodontitis reduces systemic inflammation in type 2 diabetes. *J. Clin. Periodontol.* **2020**. [CrossRef] [PubMed]
135. Baeza, M.; Morales, A.; Cisterna, C.; Cavalla, F.; Jara, G.; Isamitt, Y.; Pino, P.; Gamonal, J. Effect of periodontal treatment in patients with periodontitis and diabetes: Systematic review and meta-analysis. *J. Appl. Oral Sci. Revista FOB* **2020**, *28*, e20190248. [CrossRef] [PubMed]
136. Phillips, N.M. Does Treatment of Periodontal Disease Improve Glycemic Control in Diabetes? *Am. J. Nurs.* **2012**, *112*, 22. [CrossRef]
137. Janket, S.J.; Wightman, A.; Baird, A.E.; Van Dyke, T.E.; Jones, J.A. Does periodontal treatment improve glycemic control in diabetic patients? A meta-analysis of intervention studies. *J. Dent. Res.* **2005**, *84*, 1154–1159. [CrossRef]
138. Sastrowijoto, S.H.; Hillemans, P.; van Steenbergen, T.J.; Abraham-Inpijn, L.; de Graaff, J. Periodontal condition and microbiology of healthy and diseased periodontal pockets in type 1 diabetes mellitus patients. *J. Clin. Periodontol.* **1989**, *16*, 316–322. [CrossRef] [PubMed]
139. Tervonen, T.; Oliver, R.C.; Wolff, L.F.; Bereuter, J.; Anderson, L.; Aeppli, D.M. Prevalence of periodontal pathogens with varying metabolic control of diabetes mellitus. *J. Clin. Periodontol.* **1994**, *21*, 375–379. [CrossRef] [PubMed]
140. Kuo, L.C.; Polson, A.M.; Kang, T. Associations between periodontal diseases and systemic diseases: A review of the inter-relationships and interactions with diabetes, respiratory diseases, cardiovascular diseases and osteoporosis. *Public Health* **2008**, *122*, 417–433. [CrossRef]
141. Makiura, N.; Ojima, M.; Kou, Y.; Furuta, N.; Okahashi, N.; Shizukuishi, S.; Amano, A. Relationship of Porphyromonas gingivalis with glycemic level in patients with type 2 diabetes following periodontal treatment. *Oral Microbiol. Immunol.* **2008**, *23*, 348–351. [CrossRef] [PubMed]
142. Ojima, M.; Takeda, M.; Yoshioka, H.; Nomura, M.; Tanaka, N.; Kato, T.; Shizukuishi, S.; Amano, A. Relationship of periodontal bacterium genotypic variations with periodontitis in type 2 diabetic patients. *Diabetes Care* **2005**, *28*, 433–434. [CrossRef] [PubMed]
143. Tian, J.; Liu, C.; Zheng, X.; Jia, X.; Peng, X.; Yang, R.; Zhou, X.; Xu, X. Porphyromonas gingivalis Induces Insulin Resistance by Increasing BCAA Levels in Mice. *J. Dent. Res.* **2020**, *99*, 839–846. [CrossRef]
144. Aemaimanan, P.; Amimanan, P.; Taweechaisupapong, S. Quantification of key periodontal pathogens in insulin-dependent type 2 diabetic and non-diabetic patients with generalized chronic periodontitis. *Anaerobe* **2013**, *22*, 64–68. [CrossRef] [PubMed]
145. Arimatsu, K.; Yamada, H.; Miyazawa, H.; Minagawa, T.; Nakajima, M.; Ryder, M.I.; Gotoh, K.; Motooka, D.; Nakamura, S.; Iida, T.; et al. Oral pathobiont induces systemic inflammation and metabolic changes associated with alteration of gut microbiota. *Sci. Rep.* **2014**, *4*, 4828. [CrossRef]
146. Castrillon, C.A.; Hincapie, J.P.; Yepes, F.L.; Roldan, N.; Moreno, S.M.; Contreras, A.; Botero, J.E. Occurrence of red complex microorganisms and Aggregatibacter actinomycetemcomitans in patients with diabetes. *J. Investig. Clin. Dent.* **2015**, *6*, 25–31. [CrossRef] [PubMed]

Review

A Concise Review of Silver Diamine Fluoride on Oral Biofilm

Jingyang Zhang, Sofiya-Roksolana Got, Iris Xiaoxue Yin, Edward Chin-Man Lo and Chun-Hung Chu *

Faculty of Dentistry, The University of Hong Kong, Hong Kong SAR 999077, China; amzjy@connect.hku.hk (J.Z.); sofiya@hku.hk (S.-R.G.); irisxyin@hku.hk (I.X.Y.); hrdplcm@hku.hk (E.C.-M.L.)
* Correspondence: chchu@hku.hk; Tel.: +852-28590287

Abstract: Studies have shown that silver diamine fluoride (SDF) is an effective agent to arrest and prevent dental caries due to its mineralizing and antibacterial properties. While plenty of studies have investigated the mineralizing properties, there are few papers that have examined its antibacterial effect on oral biofilm. The objective of this study was to identify the effect of silver diamine fluoride on oral biofilm. Method: The keywords used were (silver diamine fluoride OR silver diammine fluoride OR SDF OR silver fluoride OR AgF AND biofilm OR plaque). Two reviewers screened the titles and abstracts and then retrieved the full text of the potentially eligible publications. Publications of original research investigating the effect of SDF on oral biofilm were selected for this review. Results: This review included 15 laboratory studies and six clinical studies among the 540 papers identified. The laboratory studies found that SDF could prevent bacterial adhesion to the tooth surface. SDF also inhibited the growth of cariogenic bacteria, including *Streptococcus mutans*, *Lactobacillus acidophilus*, *Streptococcus sobrinus*, *Lactobacillus rhamnosus*, *Actinomyces naeslundii*, and *Enterococcus faecalis*, thus contributing to its success in caries arrest. One clinical study reported a decrease in *Streptococcus mutans* and *Lactobacillus* sp. in arrested caries after SDF treatment, and another clinical study found that SDF inhibited the growth of periodontitis microbiota, including *Porphyromonas gingivalis*, *Tannerella forsythia*, and *Prevotella intermedia/nigrescens*. However, three clinical studies reported no significant change in the microbial diversity of the plaque on the tooth after SDF treatment. Moreover, one laboratory study and one clinical research study reported that SDF inhibited the growth of *Candida albicans*. Conclusion: Not many research studies have investigated the effects of SDF on oral biofilm, although SDF has been used as a caries-arresting agent with antibacterial properties. However, a few publications have reported that SDF prevented bacterial adhesion to the teeth, inhibited the growth of cariogenic and periodontal bacteria, and possessed antifungal properties.

Keywords: silver diamine fluoride; SDF; silver fluoride; AgF; biofilm; plaque

Citation: Zhang, J.; Got, S.-R.; Yin, I.X.; Lo, E.C.-M.; Chu, C.-H. A Concise Review of Silver Diamine Fluoride on Oral Biofilm. *Appl. Sci.* **2021**, *11*, 3232. https://doi.org/10.3390/app11073232

Academic Editor: Yoshiaki Nomura

Received: 26 February 2021
Accepted: 30 March 2021
Published: 4 April 2021

Publisher's Note: MDPI stays neutral with regard to jurisdictional claims in published maps and institutional affiliations.

1. Introduction

The oral microbiota colonizes oral biofilm on the surface of the tooth or on the mucosa within the mouth. The accumulation of oral biofilm can lead to oral diseases such as dental caries or periodontitis [1]. Oral biofilm can lead to dental caries in two ways. First, bacteria within the oral biofilm produce acid through sugar metabolism. Second, the acid causes a subsequent decrease in the environmental pH value. Both of the above are responsible for the demineralization of the tooth surface and the formation of dental caries [2]. The amount of tooth mineral and other calcium phosphates in the plaque fluid decreases rapidly after exposure to fermentable carbohydrates. Lactic acid production and a reduction in the plaque fluid volume can result in the formation of caries [3].

Acid-base-producing bacteria, in varying numbers and proportions, are the key pathogens associated with dental caries [4]. Among the suspected acid-producing bacteria, *Mutans streptococci* (MS) has been confirmed, in a systematic literature review, to play a central role in the initiation of dental caries on both enamel and root surfaces [5]. This is the case for several reasons: first, MS is the most frequently isolated species from a caries lesion; second, MS is highly acidogenic and aciduric [6,7]; and third, MS can produce surface antigens I/II and water-insoluble glucan, which promote bacterial adhesion to the tooth surface and to other bacteria [6]. For the purpose of halting the closed circle of the process, an intervention that can inhibit the bacteria that produce acid is needed.

Silver diamine fluoride (SDF) has been widely used to prevent and arrest caries, and "silver diamine fluoride" is the most common spelling/keyword for this compound in the dental literature. SDF was approved for use as a therapeutic agent in Japan over 50 years ago [8]. It has also been used for dental caries treatment in Australia and in some countries in Latin America such as Argentina and Brazil for many years [9]. The Food and Drug Administration approved the marketing of SDF in the United States for the treatment of human dentinal hypersensitivity in adults in 2014 [10].

SDF contains silver and fluoride, which form a complex with ammonia in a colorless alkaline solution [9]. It is not only a combination of silver salt, ammonium, and fluoride ions but also a mixed heavy-metal halide co-ordination complex. Ammonia can provide an alkaline environment, thus keeping the solution constant for a certain period of time [11]. In addition, silver compounds have been reported to have antimicrobial properties in the application of medicine and dentistry for decades [12]. Fluoride is also used in various forms to prevent and arrest caries due to its remineralization function [9]. Therefore, SDF has been hypothesized to contain the combined effects of silver and fluorides.

Fluoride can inhibit the production and metabolic activity of certain strains of bacteria within biofilm at low concentrations [13]. At high concentrations, fluoride inhibits the growth of cariogenic bacteria in dental plaque [14]. One mechanism of biofilm inhibition is to bind its ions to the bacterial cell constituents and to influence enzymes such as enolase and proton-extruding adenosine triphosphatase [15]. Another mechanism is the inhibition of the carbohydrate metabolism of acidogenic oral bacteria [16,17]. However, this may contribute to the development of bacterial tolerance and shorten the duration of the antibacterial effect [18]. Nonetheless, the dominant advantage of fluoride is that the re-mineralizing effect, particularly for the SDF solution, effectively arrests dental caries [19].

Several reviews have been conducted regarding the clinical use of SDF for preventing and arresting dental caries with significant effect [20,21]. However, the rationale behind the impact of SDF on biofilm remains unclear. The aim of this systematic review was to identify the effect of SDF on biofilm. The study question was as follows: What is the effect of SDF on biofilm?

2. Materials and Methods

Two independent reviewers (Zhang and Got) conducted a literature search using the four most commonly used databases: Medline, PubMed, Embase, and Scopus. Only papers published in English were included. The keywords used were (silver diamine fluoride OR silver diamine fluoride OR SDF OR silver fluoride OR AgF AND biofilm OR plaque). No publication date limitation was established during the search. The latest search was conducted on 30 October 2020.

The inclusion criteria were as follows: (1) papers investigating biofilm; (2) SDF solution at any concentration adopted as an intervention. The exclusion criteria were (1) studies with no information concerning biofilm and (2) reviews, case reports, and conference papers. In addition, a manual search of the reference lists of the selected papers and review articles was conducted. The two reviewers independently screened the papers and identified the relevant studies. When a disagreement occurred, another researcher (Chu) was consulted to achieve a consensus.

The results were summarized into three tables classified by the aim of the studies: Table 1 is for the studies investigating the effect of SDF on biofilm formed via microbiota, Table 2 is for the studies on the diversity and microbial change resulting from the application of SDF, and Table 3 is for the studies focused on the effect of SDF on fungus. Information regarding the following aspects of the background of the included studies was extracted and is presented in the summary tables: (1) authors and year; (2) materials studied;

(3) setting of the study; (4) block (or sample) adopted in the study; (5) microbiota studied; (6) assessing method for the biofilm; (7) duration of the study; and (8) findings of the study.

Table 1. Studies on the antibacterial effect of silver diamine fluoride (SDF).

Author, Year	Setting; Substrate	Microbiota	Period; Assessment	Intervention	Antibacterial Effect
Hiraishi et al., 2010	In vitro; human dentin	*E. faecalis*	15 min, 60 min; CFU	Gp1: Ca(OH)$_2$ Gp2: SDF Gp3: NaOCl Gp4: NaCl	CFU 15 min Gp2, Gp3 < Gp1, Gp4 CFU 60 min: G2, Gp3 had no *E. faecalis*
Chu et al., 2012	In vitro; human dentin	*S. mutans*, *A. naeslundii*	7 days; CFU, CLSM	Gp1: SDF Gp2: Water	CFU: Gp1 < Gp2 CLSM: Gp2 < Gp1
Mei et al., 2013a	In vitro; human dentin	*S. mutans* *L. acidophilus*	7 days; CFU, CLSM	Gp1: SDF Gp2: Water	CFU: Gp1 < Gp2 CLSM: Gp2 < Gp1
Mei et al., 2013b	In vitro; human dentin	Consortium of *S. mutans*, *S. sobrinus*, *L. acidophilus*, *L. rhamnosus*, *A. naeslund*	7 days, 14 days, 21 days; CFU, CLSM	Gp1: SDF Gp2: Water	CFU: Gp1 < Gp2 CLSM: Gp2 < Gp1
Shah et al., 2013	In vivo	*S. mutans*	BL, 3 days, 6 months, 12 months, 18 months; CFU	Gp1: SDF Gp2: NaF + CaF Gp3: APF gel	CFU: Gp1 < Gp2 Gp1 = Gp3
Savas et al., 2015	In vitro; bovine enamel	*S. mutans*	7 days; TBC, pH	Gp1: Water Gp2: SDF	Antibacterial activity: Gp2 > Gp1
Göstemeyer et al., 2017	In vitro; bovine dentine	*L. rhamnosus*	6 days; CFU	Gp1: Water Gp2: SDF	CFU: Gp2 < Gp1
Soekanto et al., 2017	In vitro	*S. mutans*, *E. faecalis*	1 day; MIC, MBC	Gp1: SDF Gp2: NSF Gp3: PPF	MIC: Gp2: 3%-*S.mutans*, 3%-*E.faecalis*; Gp3: 3%-*S.mutans*, 6%-*E.faecalis*; MBC: Gp2: 4%-*S.mutans*, 4%-*E.faecalis*; Gp3: 10%-*S.mutans*
Göstemeyer et al., 2018	In vitro; bovine dentin	*L. rhamnosus*	12 days; CFU	Gp1: SDF Gp2: CHX Gp3: N/T	CFU: No significant difference
Vinson et al., 2018	In vitro	*S. mutans*	1 day; CFU	Gp1: SDF Gp2: SDF + KI Gp3: KI	CFU: Gp3 > Gp2 > Gp1
Yu et al., 2018	In vitro; human dentin	*S. mutans*	7 days; CFU, CLSM	Gp1: SDF + NaF Gp2: SDF Gp3: NaF Gp4: Water	CFU: Gp2 < Gp1 < Gp4 < Gp3 CLSM: Gp2 > Gp1 > Gp3,Gp4
Al-Madi et al., 2019	In vitro; human dentin	*E. faecalis*	21 days; CLSM	Gp1: SDF Gp2: CHX Gp3: NaOCl	CLSM: Gp3 >Gp1 > Gp2
Wu et al., 2019	In vitro	*S. mutans*	4 days: inhibition zone; 7 days: biofilm assay	Gp1: SDF Gp2: No silver Gp3: AgNO$_3$	Inhibition zone: Gp1,Gp3 > Gp2, Antibacterial activity: Gp1 > Gp2

Table 1. *Cont.*

Author, Year	Setting; Substrate	Microbiota	Period; Assessment	Intervention	Antibacterial Effect
Sorkhdini et al., 2020	In vitro; human enamel	*S. mutans*	3 days; CFU	Gp1: SDF Gp2: SDF + KI Gp3: AgNO$_3$ Gp4: Water	CFU:Gp1 < Gp2 < Gp3 < Gp4
Rams et al., 2020	In vitro	Plaque from adults with SP	7 days; CFU, MALDI-TOF	Gp1: SDF Gp2: N/T	TVC: Gp1 < Gp 2 PLPP: Gp1 < Gp 2

Abbreviations: Gp, group; CFU, colony forming unit; CLSM, confocal laser scanning microscopy (dead-to-live ratio); NaCl, sodium chloride; BL, baseline; Ca(OH)$_2$, calcium hydroxide; NaOCl, sodium hypochlorite; TBC, total bacteria count; APF, acidulated phosphate fluoride; AHF, ammonium hexafluorosilicate; CPC, cetylpyridinium chloride; CHX, chlorhexidine; NaF, sodium fluoride; N/T, no treatment; NSF, nano silver fluoride; PPF, propolis fluoride; KI, potassium iodide; AgNO$_3$, silver nitrate; SP, severe periodontitis; TVC, total viable counts; PLPP, proportional levels of periodontal pathogens; MALDI-TOF, matrix-assisted laser desorption/ionization time-of-flight.

Table 2. Microbiota change within biofilm after silver diamine fluoride treatment.

Authors, Year	Setting; Plaque Sample	Microbiome Detecting Method; Time Points	Diversity Assessing Method or Index	Main Findings
Milgrom et al., 2018	In vivo; plaque from children	RNA sequencing; BL, follow-up after 14 to 21 days	LME4 (R core team)	*S. mutans* and *lactobacilli*: in all samples except one; Cariogenic bacteria: no significant changes; Diversity analyses: no significant differences
Mitwalli et al., 2019	In vivo; plaque from adults	16S rDNA PCR; BL, 1 month	Shannon–Weaver index, Fisher's alpha	Reduced: *Actinomyce sp.*, *P. acidifaciens*, *S. inopinata*, *Propionibacterium sp.*, *T. denticola*, *B. dentium*, *P. denticolens*, *A. israelii*; Diversity analyses: no significant difference pre- and post-intervention
Mei et al., 2020	In vivo; plaque from children	16S rRNA gene sequencing; BL, 2 weeks, 12 weeks	Shannon index	Active caries: reduced diversity 12 weeks after SDF; *S. mutans*, *Lactobacillus* sp. increased; Arrested caries: *S. mutans*, *Lactobacillus* sp. reduced; Diversity analyses: no significant change before and after 2 or 12 weeks in active caries
Liu et al., 2020	In vitro; plaque from children	16S rRNA gene sequencing; BL, 1 day, 7 days	Shannon index	Diversity analyses: no significant change in plaque; Reduction in carbohydrate transportation and metabolic functions in plaque at 1 day and 7 days post-intervention

Abbreviations: BL, baseline; LMER, linear mixed effects regression; OTU, operational taxonomic unit.

Table 3. Studies on the antifungal effect of silver diamine fluoride.

Authors, Year	Setting; Substrate	*Candida* Species	Period; Assessment	Intervention Group	Main Findings
Alshahni et al., 2020	In vitro; human dentin	*C. albicans*	3 days; CFU, CQ, Rt-PCR, SEM	Gp1: 3.8% SDF Gp2: 38% SDF Gp3: No treatment	CFU: Gp1,Gp2 < Gp3 CQ: Gp1,Gp2 < Gp3 Rt-PCR: Gp1,Gp2 < Gp3 SEM: Cell wall damage in Gp1, Gp2
Fakhruddin et al., 2020	In vitro; paper disc	*C. albicans, C. glabrata, C. parapsilosis, C. tropicalis, C. krusei, C. dubliniensis*	2 days; ZGI, Multi-PCR, MIC	Gp1: Amphotericin B Gp2: Fluoconazole Gp3: SDF complex	Gp3: anti-candidal potency; UI: Cell wall damage in Gp3; ZGI: *C. tropicalis*: Gp3 < Gp1,Gp2; *C. krusei, C. glabrata, C. albicans*: Gp3 > Gp1,Gp2

Abbreviations: CQ, colorimetric quantification; Rt-PCR, real-time PCR-based quantification; SEM, scanning electron microscopy; ZGI, zones of growth inhibition; Multi-PCR, multiplex PCR; MIC, minimal inhibitory concentration; UI, ultrastructural image.

3. Results

A total of 540 potentially relevant records were identified through the database search, and 194 were left after the removal of duplicates. A manual search of the references of these selected publications was conducted. One article that met the inclusion criteria was added to the list. After the titles and abstracts were screened, 133 records were excluded. The remaining 62 papers, all published in English, were retrieved for full-text reading. Among them, 20 articles did not report microbial analysis and 21 did not contain biofilm or microbiome information.

Among the 21 papers included, six were laboratory studies and 15 were clinical studies. The 21 included papers were published between 2010 and 2020. An SDF solution concentration of 38% was reported in 16 studies, followed by 3.8% SDF in three studies and 25% SDF, or 8.5 wt%, in the other studies. Figure 1 is a flowchart showing the search results.

3.1. Application of Silver Diamine Fluoride on Bacteria

3.1.1. Antibacterial Effect

A total of 15 studies reporting the antibacterial effects of SDF on bacteria are summarized in Table 1. Among them, 14 studies measured the antibacterial properties using monospecies bacteria such as *Lactobacillus acidophilus*, *Streptococcus sobrinus*, *Enterococcus faecalis*, and *Actinomyces naeslundii*. However, subgingival microbial biofilm specimens were used on 24 adults with severe periodontitis [22]. The majority of the included studies that explored bactericidal properties were operated in vitro on dentin or enamel samples, with no difference in the effect regarding the sample material. Only one research study was conducted in vivo [23]. The longest study duration was 21 days, whereas the shortest one was 14 min. They also revealed that biofilm treated with SDF had fewer bacteria compared with that treated with water or other interventions. However, one study did not detect any significant difference among the study groups [24]. Among the included studies, eight out of nine studies reported that SDF inhibited the growth of *Streptococcus mutans* through colony-forming unit counts. Two studies revealed that SDF also had an antibacterial function in *Lactobacillus acidophilus* and *Actinomyces naeslundii*. In addition, only one out of the two studies reported that SDF reduced the amount of *Enterococcus faecalis* [25] while the other reported reduced *Lactobacillus rhamnosus* [26]. One study reported that a 38% and a 19% SDF solution inhibited cultivable bacteria without any significant difference [27]. Only *P. micra* and *S. constellatus* belong to red- and orange-complex species recovered from SDF-treated specimens.

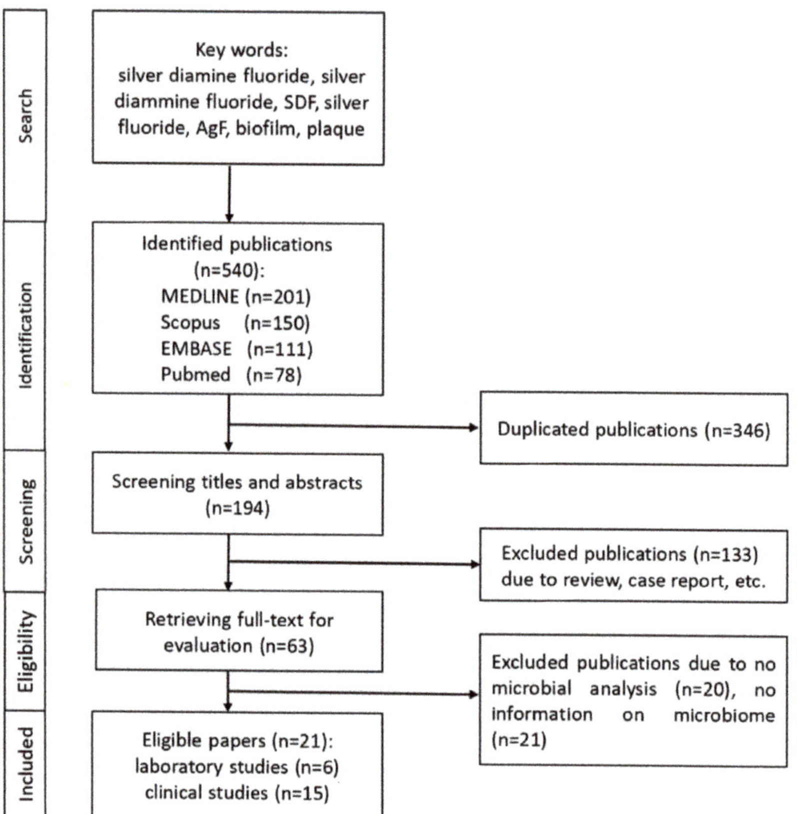

Figure 1. Flowchart of the study selection process: identification, selection, eligibility, included.

Furthermore, some studies reported that the live-to-dead ratios of *Streptococcus mutans*, *Lactobacillus acidophilus*, and *Actinomyces naeslundii* in biofilm were significantly lower after the application of SDF [23–27]. Only Al-Madi and co-workers reported that a 5.25% sodium hypochlorite solution had a significantly higher live-to-dead ratio of *Enterococcus faecalis* compared with the study's SDF group [27]. Only one study reported that the antibacterial effects of nano silver fluoride varnish and the potency of propolis fluoride varnish were comparable with a 38% SDF varnish. Another study compared the minimum inhibitory concentration and the minimum bactericidal concentration of propolis fluoride and nano silver fluoride for the inhibition of biofilm formation [28].

3.1.2. Microbiota Change in Community Diversity and/or Composition within Biofilm

A total of four studies reported on the microbiota diversity before and after SDF treatment. In terms of intervention with SDF solution, only a 38% concentration was adopted in all four studies (Table 2). The majority of the included studies that explored bactericidal properties were operated in vivo, with one study conducted in vitro; it used extracted caries teeth in an artificial mouth model [29]. Both saliva and plaque were collected in two studies [29,30], whereas only plaque was collected in the other two studies [31,32]. Two studies were conducted on preschool children, whereas one was conducted on primary school students [29], and another was conducted on adults with a mean age of 47 years [32]. The shortest study duration was 24 h, whereas the longest was 12 weeks. Among the included studies, three out of the four used the polymerase chain reaction amplification of 16S ribosomal ribonucleic acid genes and MiSeq sequencing in diversity analyses. One study used ribosomal ribonucleic acid sequencing and quantitative polymerase chain reaction quantification [31]. The Shannon index was the most frequently used index for diversity assessment.

All four studies reported no significant change in diversity. However, one study found a significant decrease in the diversity of oral bacteria plaque within active caries, but not in arrested caries, 12 weeks after SDF treatment [30]. Furthermore, the same paper reported an alteration in the species of the bacteria both before and after the SDF treatment. *Streptococcus mutans* and *Streptococcus sobrinus* increased significantly after two and 12 weeks of the SDF treatment in the plaque within active caries. Meanwhile, *Lactobacillus* sp. and *Rothia* sp. increased significantly after two weeks. Another study reported that the carbohydrate transportation and metabolic functions in plaque were significantly reduced at 24 h and one week post-intervention [29].

3.2. Application of Silver Diamine Fluoride on Fungus

Two studies explored the anti-candidal effect of SDF [16,33] (Table 3). One study of *Candida albicans* on a dentin block taken from human teeth reported that the anti-candidal effect was a dose-related property. No significant differences were found between 3.8% and 38% SDF solutions for the colony-forming unit counts, the colorimetric quantification of *Candida*, the real-time polymerase chain reaction-based quantification, and the scanning electron microscopy. The 38% SDF caused severe damage to candida cell walls, whereas the 3.8% SDF caused only partial damage. In another study, *Candida albicans*, *Candida glabrata*, *Candida parapsilosis*, *Candida tropicalis*, *Candida krusei*, and *Candida dubliniensis* were isolated from the carious lesions of preschool children. Zones of growth inhibition, a multiplex polymerase chain reaction, and minimal inhibitory concentrations (MIC50 and MIC90) were adopted for assessing the anti-candidal effect of SDF and the controls. *C. tropicalis* was more resistant to SDF based on the diameter of the inhibitory zone of the culture growth. Meanwhile, *C. krusei* and *C. glabrata* were more sensitive to SDF.

4. Discussion

This review is the first review concerning the effect of SDF applied directly to biofilm. It is based on the count of bacteria and fungi as well as the diversity within the biofilm. A large number of reviews have reported the effectiveness of SDF for preventing and arresting caries in clinical studies, and its effect was reported as extraordinary, mostly due to its antibacterial function and remineralization property [20].

Although no limitation was set regarding the publication year during the search, the earliest study found was published in 2010. This shows a new trend where research is being turned into laboratory work exploring the reason for SDF's effectiveness in preventing and arresting caries. The majority (15/21, 71%) of the included studies were published between 2016 and 2020. SDF solutions featuring different concentrations [22,34] or combined with different materials [17,35,36] were adopted to investigate their potential use for antibacterial purposes.

SDF has an antibacterial function because both the silver ions and the fluoride contained in SDF appear to have the ability to inhibit the formation of cariogenic biofilm [37]. A 38% SDF solution contains approximately 253,870 ppm silver and 44,800 ppm fluoride ions [38]. The microorganisms can be killed and silver ions can interfere with metabolic processes [12]. A three-pronged approach exists for silver ions to kill the microbiota: damage the cell wall structure of bacteria, influence metabolic processes and inhibit enzyme activities, and, finally, inhibit the replication of bacterial deoxyribonucleic acid [12]. In addition, it has been suggested that silver ions at a concentration of 20 ppm can inhibit the growth of *Streptococcus mutans* [39]. It has also been reported that the antimicrobial effect stems from the silver ions, especially at low concentrations [40]. Furthermore, the antimicrobial function of a high concentration of fluoride cannot be ignored [41]. The reason for this is that a high concentration of fluoride can influence enzymes' carbohydrate metabolism and sugar uptake and bind the bacterial cellular components, resulting in the inhibition of biofilm formation [37]. However, only one study reported no significant difference in the count of *Lactobacillus rhamnosus* among the 38% SDF group and the other groups (35% chlorhexidine varnish, 5% sodium fluoride varnish, 500 ppm sodium fluoride solution + 0.1% chlorhexidine solution, and the blank control) [26]. The possible reason for this may be that the formation of the SDF adopted in the study was a varnish rather than a solution, which indicates that the reduction in bioactive silver ions was not the same as that in the 38% SDF solution.

According to this review, even though SDF inhibits some oral microbiota species, the diversity of the biofilm does not seem to change before and after SDF application. One possible reason for this is that SDF only reduces the number of carious species, for example, *Streptococcus mutans* and *Lactobacillus*, rather than inhibiting all of the microbiota species. In the caries process, *Streptococci*, *Lactobacilli*, and *Actinomycetes* are treated as the initial bacterial invasion species. Among them, *Streptococcus mutans* is one of the most important pathogens associated with the initiation and progression of caries [42]. *Lactobacillus acidophilus* and *Lactobacillus rhamnosus* are routinely found in deep and superficial carious lesions, which indicates that they are the most abundant bacteria species [43,44]. *Actinomyces naeslundii* is related to root caries, which have the potential to invade dentinal tubules [15]. After the results of the studies were analyzed, it was found that the possible mode of action of SDF can be related to its antibacterial properties on cariogenic bacteria rather than a change in the diversity of the biofilm. The microbiota appeared to reach a new balance, but with fewer carious species [45].

According to this review, SDF also has antifungal properties. As discussed above, the bioactive form of silver in SDF in its ionized form is "Ag+". It has been reported that silver particles can inhibit the growth of *C. albicans* under high concentrations [46]. One of the possible reasons for this is that silver ions can also suppress extracellular phospholipase production, which plays a crucial role in the pathogenicity of *C. albicans* [47]. Phospholipase is associated with the development of hyphae, which plays an essential role in the biofilm's adherence and formation [48]. The blocking transformation from a yeast form to a hyphal form could stop colonization and initiation or pathogenesis. However, variations in inhibition still exist among different species in the yeast. Another possible explanation for this could be the avidity of the biological ligands of the yeast to SDF [49].

Based on the review, the most frequently used approach for assessing biofilm is still the colony-forming unit count and the live-to-dead ratio in the traditional culture method. New molecular biological technology is being adopted into assessments, usually combined with traditional ones. The majority of the studies included in this review were conducted in vitro. As the duration of the selected laboratory studies was relatively short, the long-term caries-arresting effect and the periodicity of the SDF application could not be evaluated. It was not the objective of this review to judge the quality of the studies or to discuss the limitations of each study. This should be taken into consideration when interpreting the results as well as the conclusions of this review.

This concise review was intended to provide the best evidence regarding SDF's antimicrobial effect, as presented in 21 selected papers, to obtain a definitive answer to a research question involving antibacterial function. It provides an overview and updated information about the antimicrobial effect of SDF on oral biofilm, as well as its known role in the inhibition of cariogenic bacteria. The major lines of investigation in this review involved 15 laboratory studies and six clinical studies. Some data are quite limited due to the lack of studies on certain topics such as the antibacterial effect of SDF on periodontal pathogens. As a whole, this field shows significant gaps, with only one study so far providing limited information [22]. In addition, only two published papers have shown promising results for SDF's antifungal effect against candida [16,33]. However, overall, this review provides strong evidence that SDF can, indeed, inhibit the growth of cariogenic bacteria and prevent bacterial adhesion to the tooth surface. This review—the first concise review of the effect of SDF applied directly to biofilm—can be used to guide future research in this area.

Even though SDF has been used for decades, its properties remain under-investigated. Thus far, most laboratory studies have focused on its remineralization properties rather than its antimicrobial functions within biofilm. It has been suggested that the concentrations of antibacterial agents required to inhibit biofilm are more than 100 times higher than those needed to inhibit planktonic bacteria, as biofilm is more resistant to antimicrobial agents than planktonic bacteria are [12]. More studies exploring the rationale behind these materials are still needed.

5. Conclusions

The number of publications exploring the effects of SDF on oral biofilm is limited, although SDF has been used as a caries-arresting agent with antibacterial properties. The limited publications reported that SDF prevented bacterial adhesion to teeth, inhibited the growth of cariogenic and periodontal bacteria, and possessed antifungal properties.

Author Contributions: Writing—original draft preparation, J.Z. and S.-R.G.; writing—review and editing, C.-H.C., E.C.-M.L., and I.X.Y.; supervision, C.-H.C. All authors have read and agreed to the published version of the manuscript.

Funding: This research received no external funding.

Informed Consent Statement: Not applicable.

Conflicts of Interest: The authors declare no conflict of interest.

References

1. Washio, J.; Mayanagi, G.; Takahashi, N. Challenge to metabolomics of oral biofilm. *J. Biosci.* **2010**, *52*, 225. [CrossRef]
2. Takahashi, N.; Nyvad, B. The role of bacteria in the caries process: Ecological perspectives. *J. Dent. Res.* **2011**, *90*, 294–303. [CrossRef]
3. Margolis, H.; Moreno, E. Composition and cariogenic potential of dental plaque fluid. *Crit. Rev. Oral Biol. Med.* **1994**, *5*, 1–25. [CrossRef]
4. Kleinberg, I. A mixed-bacteria ecological approach to understanding the role of the oral bacteria in dental caries causation: An alternative to Streptococcus mutans and the specific-plaque hypothesis. *Crit. Rev. Oral Biol. Med.* **2002**, *13*, 108–125. [CrossRef]
5. Tanzer, J.M.; Livingston, J.; Thompson, A.M. The microbiology of primary dental caries in humans. *J. Dent. Educ.* **2001**, *65*, 1028–1037. [CrossRef]
6. Hamada, S.; Slade, H.D. Biology, immunology, and cariogenicity of Streptococcus mutans. *Microbiol. Rev.* **1980**, *44*, 331–384. [CrossRef]
7. Loesche, W.J. Role of Streptococcus mutans in human dental decay. *Microbiol. Rev.* **1986**, *50*, 353–380. [CrossRef]
8. Yamaga, R.; Yokomizo, I. Arrestment of caries of deciduous teeth with diamine silver fluoride. *Dent. Outlook* **1969**, *33*, 1007–1013.
9. Chu, C.H.; Lo, E.C.M. Promoting caries arrest in children with silver diamine fluoride: A review. *Oral Health Prev. Dent.* **2008**, *6*, 315–321.
10. Horst, J.A. Silver fluoride as a treatment for dental caries. *Adv. Dent. Res.* **2018**, *29*, 135–140. [CrossRef]
11. Mei, M.L.; Lo, E.C.M.; Chu, C.H. Clinical use of silver diamine fluoride in dental treatment. *Compend. Contin. Educ. Dent.* **2016**, *37*, 93–98. [PubMed]
12. Peng, J.Y.; Botelho, M.; Matinlinna, J. Silver compounds used in dentistry for caries management: A review. *J. Dent.* **2012**, *40*, 531–541. [CrossRef]
13. Nassar, H.M.; Gregory, R.L. Biofilm sensitivity of seven Streptococcus mutans strains to different fluoride levels. *J. Oral Microbiol.* **2017**, *9*, 1328265. [CrossRef] [PubMed]
14. Pandit, S.; Kim, J.E.; Jung, K.H.; Chang, K.W.; Jeon, J.G. Effect of sodium fluoride on the virulence factors and composition of Streptococcus mutans biofilms. *Arch. Oral Biol.* **2011**, *56*, 643–649. [CrossRef] [PubMed]
15. Chu, C.H.; Mei, L.; Seneviratne, C.J.; Lo, E.C.M. Effects of silver diamine fluoride on dentine carious lesions induced by Streptococcus mutans and Actinomyces naeslundii biofilms. *Int. J. Paediatr. Dent.* **2012**, *22*, 2–10. [CrossRef]
16. Fakhruddin, K.S.; Egusa, H.; Ngo, H.C.; Panduwawala, C.; Pesee, S.; Venkatachalam, T.; Samaranayake, L.P. Silver diamine fluoride (SDF) used in childhood caries management has potent antifungal activity against oral Candida species. *BMC Microbiol.* **2020**, *20*, 95. [CrossRef]
17. Yu, O.Y.; Zhao, I.S.; Mei, M.L.; Lo, E.C.M.; Chu, C.H. Caries-arresting effects of silver diamine fluoride and sodium fluoride on dentine caries lesions. *J. Dent.* **2018**, *78*, 65–71. [CrossRef]
18. Lu, M.; Xiang, Z.; Gong, T.; Zhou, X.; Zhang, Z.; Tang, B.; Zeng, J.; Wang, L.; Cui, T.; Li, Y. Intrinsic fluoride tolerance regulated by a transcription factor. *J. Dent. Res.* **2020**, *99*, 1270–1278. [CrossRef]

19. Mei, M.L.; Lo, E.C.M.; Chu, C.H. Arresting dentine caries with silver diamine fluoride: What's behind it? *J. Dent. Res.* **2018**, *97*, 751–758. [CrossRef]

20. Gao, S.S.; Zhao, I.S.; Hiraishi, N.; Duangthip, D.; Mei, M.L.; Lo, E.C.M.; Chu, C.H. Clinical trials of silver diamine fluoride in arresting caries among children: A systematic review. *JDR Clin. Trans. Res.* **2016**, *1*, 201–210. [CrossRef]

21. Hendre, A.D.; Taylor, G.W.; Chávez, E.M.; Hyde, S. A systematic review of silver diamine fluoride: Effectiveness and application in older adults. *Gerodontology* **2017**, *34*, 411–419. [CrossRef] [PubMed]

22. Rams, T.E.; Sautter, J.D.; Ramírez-Martínez, G.J.; Whitaker, E.J. Antimicrobial activity of silver diamine fluoride on human periodontitis microbiota. *Gen Dent.* **2020**, *68*, 24–28.

23. Shah, S.; Bhaskar, V.; Venkataraghavan, K.; Choudhary, P.; Ganesh, M.; Trivedi, K. Efficacy of silver diamine fluoride as an antibacterial as well as antiplaque agent compared to fluoride varnish and acidulated phosphate fluoride gel: An in vivo study. *Indian J. Dent. Res.* **2013**, *24*, 575–581. [CrossRef] [PubMed]

24. Göstemeyer, G.; Kohls, A.; Paris, S.; Schwendicke, F. Root caries prevention via sodium fluoride, chlorhexidine and silver diamine fluoride in vitro. *Odontology* **2018**, *106*, 274–281. [CrossRef] [PubMed]

25. Hiraishi, N.; You, C.K.; King, N.M.; Tagami, J.; Tay, F.R. Antimicrobial efficacy of 3.8% silver diamine fluoride and its effect on root dentin. *J. Endod.* **2010**, *36*, 1026–1029. [CrossRef] [PubMed]

26. Göstemeyer, G.; Schulze, F.; Paris, S.; Schwendicke, F. Arrest of root carious lesions via sodium fluoride, chlorhexidine and silver diamine fluoride in vitro. *Materials* **2017**, *11*, 9. [CrossRef] [PubMed]

27. Al-Madi, E.M.; Al-Jamie, M.A.; Al-Owaid, N.M.; Almohaimede, A.A.; Al-Owid, A.M. Antibacterial efficacy of silver diamine fluoride as a root canal irrigant. *Clin. Exp. Dent. Res.* **2019**, *5*, 551–556. [CrossRef]

28. Soekanto, S.A.; Marpaung, L.J.; Himmatushohwah Djais, A.A.; Rina, R. Efficacy of propolis fluoride and nano silver fluoride for inhibition of streptococcus mutans and enterococcus faecalis biofilm formation. *Int. J. App. Pharm.* **2017**, *9*, 51–54. [CrossRef]

29. Liu, B.Y.; Mei, L.; Chu, C.H.; Lo, E.C.M. Effect of silver fluoride in preventing the formation of artificial dentinal caries lesions in vitro. *Chin. J. Dent. Res.* **2019**, *22*, 273–280.

30. Mei, M.L.; Yan, Z.; Duangthip, D.; Niu, J.Y.; Yu, O.Y.; You, M.; Lo, E.C.M.; Chu, C.H. Effect of silver diamine fluoride on plaque microbiome in children. *J. Dent.* **2020**, *102*, 103479. [CrossRef]

31. Milgrom, P.; Horst, J.A.; Ludwig, S.; Rothen, M.; Chaffee, B.W.; Lyalina, S.; Pollard, K.S.; DeRisi, J.L.; Mancl, L. Topical silver diamine fluoride for dental caries arrest in preschool children: A randomized controlled trial and microbiological analysis of caries associated microbes and resistance gene expression. *J. Dent.* **2018**, *68*, 72–78. [CrossRef]

32. Mitwalli, H.; Mourao, M.D.A.; Dennison, J.; Yaman, P.; Paster, B.J.; Fontana, M. Effect of silver diamine fluoride treatment on microbial profiles of plaque biofilms from root/cervical caries lesions. *Caries Res.* **2019**, *53*, 555–566. [CrossRef] [PubMed]

33. Alshahni, R.Z.; Alshahni, M.M.; Hiraishi, N.; Makimura, K.; Tagami, J. Effect of silver diamine fluoride on reducing candida albicans adhesion on dentine. *Mycopathologia* **2020**, *185*, 691–698. [PubMed]

34. Wu, L.; Gareiss, S.K.; Morrow, B.R.; Babu, J.P.; Hottel, T.; Garcia-Godoy, F.; Li, F.; Hong, L. Antibacterial properties of silver-loaded gelatin sponges prepared with silver diamine fluoride. *Am. J. Dent.* **2019**, *32*, 276–280.

35. Vinson, L.A.; Gilbert, P.R.; Sanders, B.J.; Moser, E.; Gregory, R.L. Silver diamine fluoride and potassium iodide disruption of in vitro streptococcus mutans biofilm. *J. Dent. Child.* **2018**, *85*, 120–124.

36. Sorkhdini, P.; Gregory, R.L.; Crystal, Y.O.; Tang, Q.; Lippert, F. Effectiveness of in vitro primary coronal caries prevention with silver diamine fluoride—Chemical vs biofilm models. *J Dent.* **2020**, *99*, 103418. [CrossRef]

37. Mei, M.L.; Li, Q.L.; Chu, C.H.; Lo, E.C.M.; Samaranayake, L.P. Antibacterial effects of silver diamine fluoride on multi-species cariogenic biofilm on caries. *Ann. Clin. Microbiol. Antimicrob.* **2013**, *12*, 4. [CrossRef] [PubMed]

38. Lou, Y.; Botelho, M.G.; Darvel, B.W. Reaction of silver diamine fluoride with hydroxyapatite and protein. *J. Dent.* **2011**, *39*, 612–618. [CrossRef] [PubMed]

39. Knight, G.M.; McIntyre, J.M.; Craig, G.G.; Mulyani Zilm, P.S.; Gully, N.J. Inability to form a biofilm of Streptococcus mutans on silver fluoride-and potassium iodide-treated demineralized dentin. *Quintessence Int.* **2009**, *40*, 155–161.

40. Lou, Y.; Darvell, B.W.; Botelho, M.G. Antibacterial effect of silver diamine fluoride on cariogenic organism. *J. Contemp. Dent. Pract.* **2018**, *19*, 591–598.

41. Pandit, S.; Kim, G.; Lee, M.; Jeon, J. Evaluation of streptococcus mutans biofilms formed on fluoride releasing and non fluoride releasing resin composites. *J. Dent.* **2011**, *39*, 780–787. [CrossRef] [PubMed]

42. Esberg, A.; Sheng, N.; Mårell, L.; Claesson, R.; Persson, K.; Borén, T.; Strömberg, N. Streptococcus mutans adhesin biotypes that match and predict individual caries development. *EBioMedicine* **2017**, *24*, 205–215. [CrossRef] [PubMed]

43. Ahmed, A.; Dachang, W.; Lei, Z.; Jianjun, L.; Juanjuan, Q.; Yi, X. Effect of Lactobacillus species on Streptococcus mutans biofilm formation. *Pak. J. Pharm. Sci.* **2014**, *27*, 1523–1528. [PubMed]

44. Zhu, B.; Li, J.Y.; Hao, Y.Q.; Zhou, X.D. Establishment and evaluation of the in vitro dynamic biofilm model. *Shanghai Kou Qiang Yi Xue.* **2010**, *19*, 60–65.

45. Takahashi, N.; Nyvad, B. Caries ecology revisited: Microbial dynamics and the caries process. *Caries Res.* **2008**, *42*, 409–418. [CrossRef] [PubMed]

46. Pereira, C.A.; Costa, A.C.; Silva, M.P.; Back-Brito, G.N.; Jorge, A.O. Candida albicans and virulence factors that increases it pathogenicity. The battle against microbial pathogens: Basic science, technological advances and educational programs; microbiology series, vol. 2. *Badajoz FORMATEX Res. Cent.* **2015**, *61*, 1.

47. Modrzewska, B.; Kurnatowski, P. Adherence of Candida sp. to host tissues and cells as one of its pathogenicity features. *Ann. Parasitol.* **2015**, *61*, 1.
48. Carradori, S.; Chimenti, P.; Fazaari, M.; Granese, A.; Angiolella, L. Antimicrobial activity, synergism and inhibition of germ tube formation by crocus sativus-derived compounds against Candida spp. *J. Enzyme Inhib. Med. Chem.* **2016**, *31*, 189–193. [CrossRef]
49. Xiao, J.; Huang, X.; Alkhers, N.; Alzamil, H.; Alzoubi, S.; Wu, T.T.; Castillo, D.A.; Campbell, F.; Davis, J.; Hrzog, K. Candida albicans and early childhood caries: A systematic review and meta-analysis. *Caries Res.* **2018**, *52*, 102–112. [CrossRef]

Article

The Oral Microbiome in Children with Black Stained Tooth

Ji Young Hwang [1], Hyo-Seol Lee [2], Jaehyuk Choi [3], Ok Hyung Nam [2], Mi Sun Kim [2,4] and Sung Chul Choi [2,*]

[1] Department of Dentistry, Graduate School, Kyung Hee University, Seoul 02447, Korea; berryry@hanmail.com
[2] Department of Pediatric Dentistry, School of Dentistry, Kyung Hee University, Seoul 02447, Korea; stberryfield@gmail.com (H.-S.L.); pedokhyung@khu.ac.kr (O.H.N.); pedokms@khu.ac.kr (M.S.K.)
[3] Division of Life Sciences, College of Life Sciences and Bioengineering, Incheon National University, Incheon 22012, Korea; jaehyukc@inu.ac.kr
[4] Department of Pediatric Dentistry, Kyung Hee University Hospital at Gangdong, Seoul 02447, Korea
* Correspondence: pedochoi@khu.ac.kr; Tel.: +82-2-958-9371; Fax: + 82-2-965-7247

Received: 25 September 2020; Accepted: 9 November 2020; Published: 13 November 2020

Abstract: Black stain (BS) is a characteristic extrinsic discoloration, which occurs along the third cervical line of the buccal and/or lingual surfaces of teeth, particularly in the primary dentition of humans. BS is produced by oral bacteria and byproducts of saliva, but there is a controversy about related bacteria. The aim of this study was to identify the oral microbiome in tooth BS using pyrosequencing. It was hypothesized that the oral microbiome of BS in children might be related to black-pigment producing bacteria. Supragingival dental plaque was obtained from six children (mean 8.1 years) with BS and four children (mean 8.3 years) without BS. The bacterial metagenome was obtained by pyrosequencing. The BS group contained 348 operative taxonomic units (OTUs), whereas the control group had 293 OTUs. Microbial abundance and diversity were significantly higher in the BS group ($p < 0.05$). In the heatmap, the correlation between samples was the same as the BS scale. At the genus level, six genera—*Abiotrophia, Eikenella, Granulicatella, Neisseria, Porphyromonus* and *Streptococcus*—were significantly different between the two groups ($p < 0.05$). We suggested that compositional changes in the oral microbiome are essential, and several species in the genus *Neisseria, Porphyromonas* and *Streptococcus* may be major contributors for BS formation. Although the number of subjects was relatively limited, our study is the first species-level analysis of pyrosequencing data in BS formation.

Keywords: black stain; oral microbiota; pyrosequencing; *Neisseria*; *Porphyromonas*; *Streptococcus*

---

1. Introduction

Black stain (BS) is a characteristic, extrinsic dental discoloration that occurs on the cervical third of the tooth and follows the contour of the gingival margin, particularly in primary dentition [1]. BS is thought to be a special form of dental plaque related to calcification because it contains an insoluble ferric salt, which is likely ferric sulfide, and has high calcium and phosphate contents [2]. The ferric sulfide may be formed through a reaction between hydrogen sulfide produced by bacterial action and iron in saliva or gingival exudates.

BS occurs at any age, but there is a higher prevalence in childhood that decreases in adolescence and adulthood [3]. Studies have shown no difference in occurrence between the sexes [1]. Researchers have been interested in BS because of its correlation with the low caries experience [2,4]. However, a recent study of more than 3000 people found no association between dental caries and BS [5].

Many studies have examined BS-related bacteria, but the causative bacteria is still not clear [1,3]. Early ultrastructural examinations of BS demonstrated that the black material is a ferric salt, probably ferric sulfide, formed by the reaction between sulfide produced by bacterial action and iron in the saliva or gingival exudate [2,3]. The most commonly cultivated bacteria from BS teeth were *Actinomycetes* with the detection of rarely pigmented Gram-negative rods [6]. Black-pigmented bacteria were found primarily in extrinsic tooth stains [6,7]. Among black-pigmented bacteria, *Prevotella intermedia* and *Prevotella nigrescens* are the ones most frequently found in the oral cavity and associated with BS [3]. Soukos et al. [6] reported that *Porphyromonas gingivalis*, *P. intermedia* and *P. nigrescens* were related to BS using DNA–DNA hybridization for 40 taxa, whereas Saba et al. [7] suggested that *P. gingivalis* and *Prevotella melaninogenica* were absent and *Actinomycetes* sp. and *Aggregatibacter actinomycetemcomitans* were present in BS by means of polymerase chain reaction (PCR) and electrophoresis gel on the agarose medium for 200 people. This conflict was partly caused by limitations in the analysis methods. As a result, more advanced analysis methods are needed.

Recent advances in sequencing technologies, such as 454 pyrosequencing, have provided a deep understanding of the metagenomic diversity of the human oral microbiome [8]. Pyrosequencing can identify bacterial sequences and their amounts to give both qualitative and quantitative information of the oral microbiome, including uncultivable microorganisms [8]. For example, high-throughput pyrosequencing has revealed that dental plaques pooled from adults contain approximately 10,000 microbial phylotypes, with the ultimate diversity of oral microbiomes being estimated to contain approximately 25,000 phylotypes [9].

The aim of this study was to identify the oral microbiome in BS using pyrosequencing and to test our hypothesis that the oral microbiome of BS in children might be related to black-pigmented bacteria.

2. Material and Methods

2.1. Sampling

This study was approved by the Institutional Review Board of the School of Dentistry, Kyung Hee University (KHD-IRB-1509-5). Ten children were recruited as study subjects. Six of the children had black stained teeth while four children maintained good oral hygiene and had no black stained teeth. Oral examinations were performed on all the children, and decayed-missing-filled teeth (DMFT) indices were estimated. The BS scale was estimated using the Shourie method [10]. The criteria for classifying BS were as follows: 1; no line, 2; incomplete coalescence of pigmented spots and 3; continuous line formed by pigmented spots. After classification, subjects with parental consent were instructed not to brush their teeth for 24 h before sampling. Sampling was performed in the morning from the subjects, who were instructed to skip breakfast. Sterilized dental explorers were used to collect supragingival plaque on the buccal and lingual surfaces of all teeth. The samples were collected in 1 mL of sterile phosphate-buffered saline and stored at −80 °C after liquid nitrogen treatment.

2.2. Genomic DNA Extraction

The following molecular and sequencing procedures were performed in an outside outsourcing research laboratory. Genomic DNA (gDNA) was isolated from the dental plaque samples using a DNA extraction kit for bacteria (iNtRON Biotechnology, Seongnam, Korea) following the manufacturer's protocol. The collected samples were treated with lysozyme at 37 °C for 15 min and lysed with a buffer containing proteinase K and RNase A at 65 °C for 15 min. The lysates were mixed with a binding buffer and then passed through resin columns for purification. The harvested g DNA from the samples was quantified using a NanoDrop ND-1000 spectrophotometer (Thermo Scientific, Waltham, MA, USA).

2.3. PCR Amplification

PCR was performed to amplify the V1 to V3 region of the 16S rRNA gene from the samples' gDNA. 9F (5′-AGAGTTTGATCMTGGCTCAG-3′) and 541R (5′-ATTACCGCGGCTGCTGG-3′) primers were used in the amplification. The amplifications were carried out under the following conditions: initial denaturation at 94 °C for 5 min, followed by 30 cycles of denaturation at 94 °C for 30 s, primer annealing at 55 °C for 30 s and extension at 72 °C for 30 s with a final extension phase at 72 °C for 5 min. The PCR products were visualized under a Gel Doc system (BioRad, Hercules, CA, USA) after electrophoresis on a 2% agarose gel. The amplified products were cleaned with the QIAquick PCR purification kit (Qiagen, Valencia, CA, USA).

2.4. DNA Pyrosequencing

Equal concentrations of the PCR products were pooled together and cleaned with Agencourt Amperes magnetic purification beads (Danvers, MA, USA). The quality and size of the pooled products were assessed on an Agilent Bioanalyzer 2100 (Palo Alto, CA, USA) using a DNA 7500 chip. The pooled products were used for emulsion PCR and then loaded onto Picotiter plates for high-throughput pyrosequencing. The sequencing was performed on the GS FLX Plus system (Roche, Branford, CT, USA) according to the manufacturer's instructions at ChunLab, Inc. (Seoul, Korea).

2.5. Pyrosequencing Data Analysis

Obtained raw data were categorized into samples according to the barcode information. The barcode, linker and primers were trimmed off from the reads and reads showing ambiguous nucleotides ($2\leq$), a low-quality score (<25) or a short length (<300 bp) were removed. Potential chimera sequences detected by the Bellerophon method were filtered. The final reads were compared with sequence and taxonomic information acquired from the EzTaxon-e database. This database has curated 16S rRNA gene sequences of species and phylotypes of either cultured or uncultured entries in the GenBank database with taxonomy information. The read sequences were queried with BLASTN, and a maximum of five best-hit sequences were compared by pairwise alignment to determine species or phylotypes. Operational taxonomic units (OTUs) were obtained at a 97% similarity level by the CL community program (ChunLab, Inc., Seoul, Korea) after comparing with the EZ taxon database. Alpha diversity was determined using the "phyloseq" R package. A heatmap analysis was performed based on microbial abundance in the "heatplus" R package. Principal component analysis (PCA) was performed using "devtools" and "ggbiplot" R packages.

2.6. Statistical Analysis

In clinical findings, dmft indices between the BS and control groups were statistically analyzed by Mann–Whitney test ($p = 0.05$), and the association between BS formation and caries development were statistically analyzed by Chi-squared test ($p = 0.05$). After the pyrosequencing, the diversity of the oral microbiome of the subjects was statistically analyzed by the t-test ($p = 0.05$). In the heatmap, the Bray–Curtis method was used for calculating cluster dissimilarity in the clustering analysis ($p = 0.05$). The abundance of the predominant bacterial groups was analyzed by t-test ($p = 0.05$). Jonckheere's trend test was used for the trend analysis ($p = 0.05$).

3. Results

3.1. Clinical Findings

Children with BS (BS group, n = 6) whose ages ranged from 4 to 11 (mean of 8.1 years) were included in this study. The control (non-BS) group (n = 4) had a mean age of 8.3 (range 6–9) years. The BS group exhibited a mean BS score of 1.5 (BS scale 1 = 3 and BS scale 2 = 3), zero DMFT index and a 3.8 average dmft index. By contrast, four control patients showed zero BS scale, zero DMFT index and a 1.75 average dmft index. The dmft indices between the BS and control groups were not significantly different ($p > 0.05$). No significant association between BS formation and caries development was observed ($p > 0.05$).

3.2. Diversity of the Oral Microbiome

A total of 10 samples were sequenced, and 60,371 reads in total were obtained. Of all the reads, 39,013 reads (64.6%) passed quality filtering (low-quality sequence reads and a minimum length of 300 bp) and remained. Both the BS and control groups contained 28,142 and 10,871 reads with an average length of 472 and 465 bp, respectively (Supplementary Table S1). The microbial abundance based on the read number was significantly higher in the BS group than the control group (Figure 1A, $p < 0.05$). OTUs were identified by pairwise alignment to the EzTaxon database at a 3% distance. Totals of 15 phyla, 28 classes, 45 orders, 80 families, 145 genera and 423 OTUs (species) were identified in these samples. The BS group contained 348 OTUs in total (without duplication), whereas the control group had 293 OTUs. The alpha diversity of the microbiome was estimated by Simpson and Shannon indices, indicating that the diversity in the BS group was significantly higher than in the control group (Figure 1B,C, $p < 0.05$).

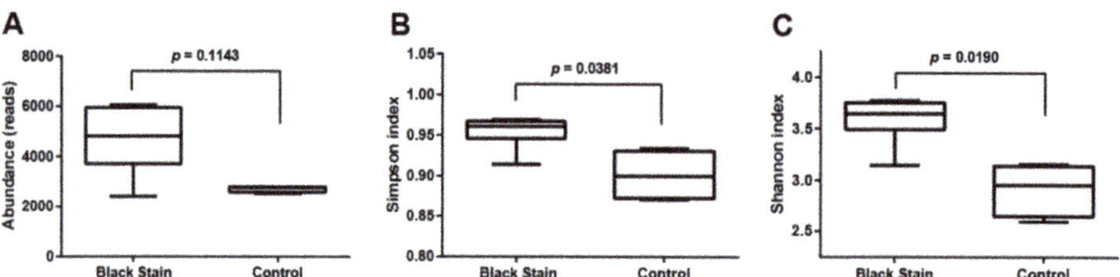

Figure 1. Diversity of the oral microbiome of the subjects. Abundance (**A**), Simpson index (**B**) and Shannon index (**C**) were compared between the black stain and control groups. Significance was determined by *t*-test ($p = 0.05$).

3.3. Different Genera and Species between BS and Control Group

Heatmap analysis was employed to see overall patterns of microbial abundance at different levels of taxonomy. At all levels from species to order, the control group was isolated from the BS group (Figure 2), suggesting that the microbial composition was quite different between the two groups. The only exception was the sample "C12", which formed its own lineage within the BS group. Since the diversity pattern of C12 was much different from the other samples in the control group, we excluded the sample from further statistical analysis. To test the hypothesis that changes in microbial diversity affected BS formation, the abundance (read number) of each OTU was compared between the groups using the Mann–Whitney U test. At the genus level, six genera—*Abiotrophia, Eikenella, Granulicatella, Neisseria, Porphyromonas* and *Streptococcus*—were significantly different between the two groups ($p < 0.05$). *Neisseria, Porphyromonas* and *Streptococcus* contained 11, 14 and 31 species within them, respectively (Figure 3). All six genera were more abundant in the BS group than the control group. In addition, the genus *Selenomonas* exhibited the opposite trend and was more abundant in the control group ($p = 0.053$). At the species level, 19 species were identified as significantly different between the two groups ($p < 0.05$, Table 1). The number of each sample represented the identified OTUs. Except for *Selenomonas noxia, Corynebacterium_uc* and *JQ454562_s* (*Prevotella*), all the species were higher in the BS group than the control group. Twelve out of 19 species were included in the genus showing significance.

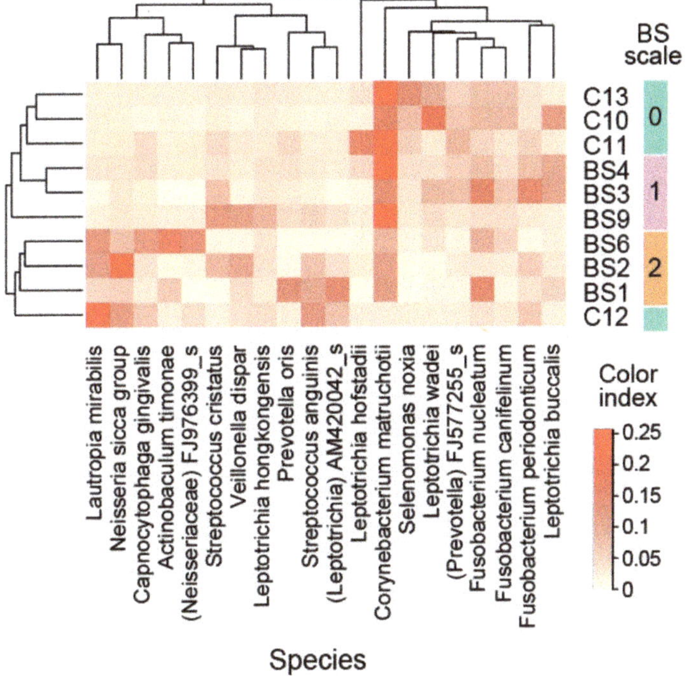

Figure 2. Heatmap of the oral microbiome relative abundance. The species having more than 5% of the total reads are shown. The Bray–Curtis method was used for calculating cluster dissimilarity in the clustering analysis (*p* = 0.05, significantly different). B: black stain group, C: control group.

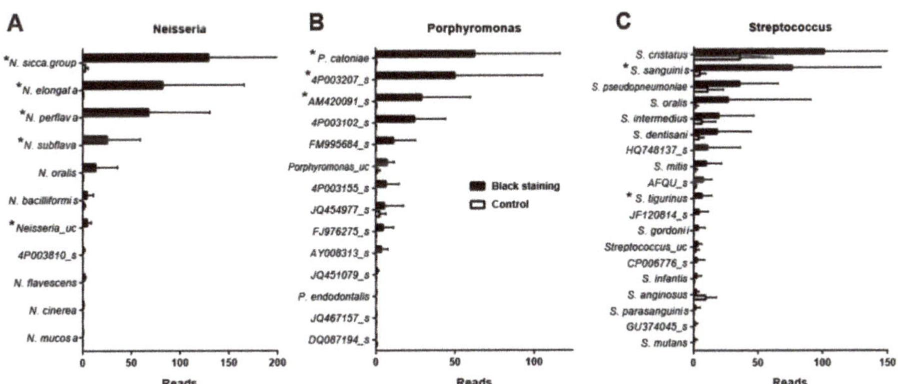

Figure 3. Abundance of the predominant bacterial groups between the black staining and control groups. Genus *Neisseria* (**A**), genus *Porphyromonas* (**B**), *Streptococcus* (**C**). Asterisks indicate species significantly different between the two groups (*p* < 0.05).

Table 1. Species showing a significant difference between the black stain and control groups ($p < 0.05$). The number of each sample representing the identified operative taxonomic units (OTUs).

Taxonomy		Black Stain Group							Control Group				p Value
Genus	Species	B1	B2	B6	B3	B4	B9	Ave.	C10	C11	C13	Ave.	
Abiotrophia	Ab. defectiva	25	122	7	7	23	2	31.0	0	0	0	0.0	0.025
Actinomyces	Ac. naeslundii	44	14	55	53	5	16	31.2	1	1	2	1.3	0.028
Cardiobacterium	C. hominis	20	3	33	53	24	4	22.8	0	1	3	1.3	0.038
Eikenella	E. corrodens	8	19	14	21	101	3	27.7	0	1	2	1.0	0.028
Granulicatella	G. adiacens	10	67	22	4	33	1	22.8	0	1	1	0.7	0.049
Kingella	JQ455165_s	6	4	10	1	19	32	12.0	0	0	0	0.0	0.026
Leptotrichia	L. hongkongensis	51	55	68	36	84	143	72.8	21	18	25	21.3	0.028
Neisseria	N. elongata	34	97	35	70	242	16	82.3	1	0	1	0.7	0.028
Neisseria	N. perflava	5	21	121	121	132	6	67.7	0	0	0	0.0	0.025
Neisseria	N. sicca group	107	360	67	32	194	15	129.2	6	2	0	2.7	0.028
Neisseria	N. subflava	1	88	6	13	41	1	25.0	0	0	0	0.0	0.025
Porphyromonas	AM420091_s	2	33	40	84	3	10	28.7	0	1	0	0.3	0.028
Porphyromonas	P. catoniae	19	97	73	38	147	1	62.5	1	0	0	0.3	0.037
Porphyromonas	4P003207_s	6	35	99	138	16	5	49.8	0	0	0	0.0	0.026
Selenomonas_g1	Se. noxia	0	5	63	37	41	7	25.5	126	85	404	205.0	0.028
Streptococcus	St. sanguinis	195	116	43	26	13	63	76.0	3	10	1	4.7	0.028

Species where all the samples had less than 10 reads on average are not included.

3.4. Trend Analysis among the Three BS Scale Groups

In the heatmap at the species level, clusters based on microbial abundance were correlated to BS scale severity (Figure 2). For example, the samples "B1", "B2" and "B9", which had been determined as having "BS scale 2" severity according to Shori's method, were clustered together in similarity (Figure 2). A similar correlation was observed at all taxonomical levels, suggesting that BS scale severity may be correlated with the bacterial composition (Figure 2). Therefore, a dose-dependent relationship between the BS scale and abundance was examined using a nonparametric Jonckheere's trend test in the "clinfun" R package. A total of 47 species showed a trend among the BS scale clusters: 39 species exhibited an "increasing" trend when the BS scale increased ($p < 0.05$), whereas eight species were found to exhibit a "decreasing" tendency ($p < 0.05$). Principal component analysis (PCA) for these selected species revealed that species showing an "increasing" trend contributed to an increase in BS scale severity (Figure 4). This microbial community may play an essential role in the formation of BS. The most abundant genera were *Neisseria*, *Porphyromonas*, *Streptococcus* and *Capnocytophaga*. In contrast, the "decreasing" population, which showed abundance in the control group, consisted of *Corynebacterium*, *Prevotella*, *Selenomonas*, *Capnocytophaga* and *Leptotrichia*. Four genera were found in both groups: *Capnocytophaga*, *Corynebacterium*, *Leptotrichia* and *Prevotella*.

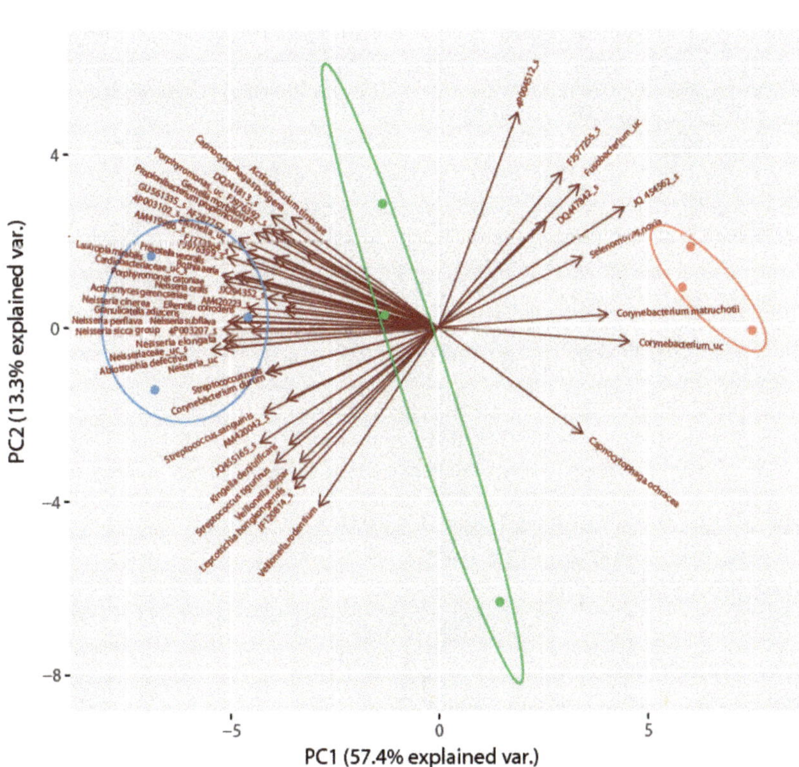

Figure 4. Principle component analysis of the selected species showing an increasing or decreasing trend according to the severity of the black stain (BS) scale. Approximately 71% of the total variance among the individual samples was explained by the first two components.

4. Discussion

Conventional BS therapy has been focused on the mechanical removal of black pigments themselves because the exact mechanism of BS prevalence had yet to be identified [1]. In this study, we used pyrosequencing analysis of supragingival dental plaque of children with and without BS to understand the microbiological aspects of BS, which may lead to better prevention and treatment for BS. The genera *Porphyromonas*, *Streptococcus* and *Neisseria*, were abundant in the microbiome of children with BS, while the genera *Corynebacterium*, *Prevotella* and *Selenomonas_g1* were abundant in children without BS in this study. This study suggests that dynamics in the oral microbiome may cause the formation of BS in children, and the microbial population in the above genera would play important roles in changing microbial dynamics.

The genus *Porphyromonas* was significantly abundant in the BS group (Figure 2). All the species belonging to this genus exhibited a similar pattern in abundance (Figure 3). Four *Porphylromonas* sp. showed a trend to increase as the BS scales increased (Figure 4). *Porphyromonas* is an obligate anaerobic bacterium that produces a black pigment called porphyrin [11]. This pigment can protect the bacterial cell from oxidative stress by reducing oxygen. *P. gingivalis* is one of the most well-known species that produces black pigments [12] and is implicated in the progression of periodontal disease [4]. However, this species was not one of the pathogens observed in this study. *Porphyromonas catoniae* is the most common species in this genus (Figure 3A). It is frequently found in the oral microbiota of healthy children [9] and is known to be a pioneer bacterium that is established in the mouths of babies [13]. In addition, it is a non-pigmented species [14]. However, it is possible that this species may play an opportunistic role in the formation of BS because this species exhibits activities of diverse cellular enzymes like trypsin, which is important in bacterial colonization [13]. The other *Porphyromonas* spp. may contribute to the accumulation of black pigments.

The genus *Streptococcus* was also observed abundantly in the BS group (Figures 2 and 3B). *S. cristatus*, *S. sanguinis*, *S. oralis*, *S. mitis* and *S. gordonii* are known as oral biofilm bacteria that colonize the tooth surface [15,16]. Since these species are normal tooth microbiome and do not produce black pigments, they may contribute to the formation of microbial beds on teeth and interact with other species producing black pigments. Although the genus *Streptococcus* makes up 10% of the total microbiome, *S. mutans* was detected in only one subject (BS2, five reads). This rare detection of *S. mutans* is consistent with a recent report of BS patients where *S. mutans* was not found. This may be due to the antagonistic effect of the predisposed *Streptococcus* spp. *S. sanguinis* and *S. gordonii* that are known to inhibit the growth of *S. mutans* by producing hydrogen peroxide [17]. A recent study using pyrosequencing suggests that the genus *Streptococcus* was most abundant but showed no difference between the groups [18]. Another pyrosequencing study reported that this genus was observed frequently in subjects having both caries and pigments, but not in subjects having only pigments [19]. Interestingly, a recent microbiome study showed that the population of *P. catoniae* was decreased in children with caries and increased in children with healthy oral status [9]. Since the *P. catoniae* population was found to be increased in our BS group, it may have a negative influence on caries development (Figure 3A).

Veillonella dispar and *V. rodentium* were identified as positive candidates for BS formation (Figure 4). *V. dispar* produces hydrogen sulfide (H_2S) from $_L$-cysteine [20]. The genus *Actinomyces* can produce H_2S and has been known to be a strong candidate for BS formation. However, *Actinomyces gerencseriae* was detected in this study, and there is no report about the H_2S production of this species.

Microbial composition is continuously affected by the environment of the oral cavity, and each population interacts with others in the same community. Therefore, a decrease in population density may result in the breakdown of one community, presenting the opportunity for a new population or community to thrive. We hypothesized that different microbial compositions would appear in different BS scales. The BS scale indicates different environmental factors affecting microbial composition. In that hypothesis, our trend analysis exhibited positive or negative candidates in BS formation. In Figure 4, some species decreased as BS scales increased, and the others were increased in proportion to the BS scale. These species may play core roles in BS formation. However, their exact roles must be identified in further studies.

Recently, two pyrosequencing studies were performed to analyze the microbiota of BS [18,19]. One was performed for 25 subjects [18], while the other study consisted of 111 subjects [19]. Li et al. [18] reported that the genera of *Actinomyces, Tannerella, Treponema, Corynebacterium, Cardiobacterium* and *Hemophilus* were more abundant in children with BS. In another study, the genera *Leptotrichia* and *Fusobacterium* contributed to BS formation, while the genera *Streptococcus* and *Mogibacterium* were important in the formation of both caries and pigments. Interestingly, three studies, including ours, did not reveal a strong consistency between them. Different ethnic backgrounds and dietary styles may provide confounding effects, suggesting that a more careful sampling design is required for this kind of study.

The limitation of this study is the small sample size. Larger sample size is needed for safe experiments and the potential confounding factor that the pyrosequencing may present. Further research will be designed by reflecting this limitation. In addition, it is meaningful to compare the study with existing healthy subjects in place of the small control group. In the study of healthy children, Proteobacteria was abundant in the group without dental caries, and Firmicutes and Bacteroides were abundant in the group with dental caries [21]. A study of healthy elderly people found that Firmicutes and Veillonella were abundant at the phyla level [22,23]. This was markedly different from the BS group in this study, in which Neisseria, Porphyromonas and Streptococcus were significant.

5. Conclusions

In conclusion, this study investigated the composition of the oral microbiome of supragingival plaque in relation to BS formation. Although the number of subjects was relatively limited, our study is the first species-level analysis of the pyrosequencing data in BS formation. We highlighted that compositional changes in the oral microbiome are essential, and several species in the genus *Neisseria, Porphyromonas* and *Streptococcus* may be major contributors to BS formation. The proposed species information will provide a foundation for understanding the complex ecology of BS and for identifying the causal agent for BS formation.

Supplementary Materials: The following are available online at http://www.mdpi.com/2076-3417/10/22/8054/s1.

Author Contributions: Conceptualization, J.Y.H. and S.C.C.; methodology, H.-S.L.; software, J.C.; validation, J.C.; formal analysis, O.H.N.; investigation, S.C.C.; resources, M.S.K.; data duration, H.-S.L.; writing—original draft preparation, J.Y.H. and S.C.C.; writing—review and editing, H.-S.L. and S.C.C.; supervision, project administration, funding acquisition, S.C.C. All authors have read and agreed to the published version of the manuscript.

Funding: This research was supported by the Basic Science Research Program of the National Research Foundation of Korea (NRF), funded by the Ministry of Education, Science and Technology (NRF-2016R1C1B1015005).

Acknowledgments: We thank the children that participated in this study.

Conflicts of Interest: The authors declare no conflict of interest.

References

1. Zyla, T.; Kawala, B.; Antoszewska-Smith, J.; Kawala, M. Black stain and dental caries: A review of the literature. *Biomed. Res. Int.* **2015**, *2015*, 469392. [CrossRef] [PubMed]
2. Reid, J.S.; Beeley, J.A.; MacDonald, D.G. Investigations into black extrinsic tooth stain. *J. Dent. Res.* **1977**, *56*, 895–899. [CrossRef] [PubMed]
3. Slots, J. The microflora of black stain on human primary teeth. *Scand. J. Dent. Res.* **1974**, *82*, 484–490. [CrossRef] [PubMed]
4. Ximenez-Fyvie, L.A.; Haffajee, A.D.; Socransky, S.S. Microbial composition of supra- and subgingival plaque in subjects with adult periodontitis. *J. Clin. Periodontol.* **2000**, *27*, 722–732. [CrossRef]

5. Garcia Martin, J.M.; Gonzalez Garcia, M.; Seoane Leston, J.; Llorente Pendas, S.; Diaz Martin, J.J.; Garcia-Pola, M.J. Prevalence of black stain and associated risk factors in preschool Spanish children. *Pediatr. Int.* **2013**, *55*, 355–359. [CrossRef]

6. Soukos, N.S.; Som, S.; Abernethy, A.D.; Ruggiero, K.; Dunham, J.; Lee, C.; Doukas, A.G.; Goodson, J.M. Phototargeting oral black-pigmented bacteria. *Antimicrob. Agents Chemother.* **2005**, *49*, 1391–1396. [CrossRef]

7. Saba, C.; Solidani, M.; Berlutti, F.; Vestri, A.; Ottolenghi, L.; Polimeni, A. Black stains in the mixed dentition: A PCR microbiological study of the etiopathogenic bacteria. *J. Clin. Pediatr. Dent.* **2006**, *30*, 219–224. [CrossRef]

8. Keijser, B.J.; Zaura, E.; Huse, S.M.; van der Vossen, J.M.; Schuren, F.H.; Montijn, R.C.; Ten Cate, J.M.; Crielaard, W. Pyrosequencing analysis of the oral microflora of healthy adults. *J. Dent. Res.* **2008**, *87*, 1016–1020. [CrossRef]

9. Crielaard, W.; Zaura, E.; Schuller, A.A.; Huse, S.M.; Montijn, R.C.; Keijser, B.J. Exploring the oral microbiota of children at various developmental stages of their dentition in the relation to their oral health. *BMC Med. Genom.* **2011**, *4*, 22. [CrossRef]

10. Shourie, K.L. Mesenteric line or pigmented plaque; a sign of comparative freedom from caries. *J. Am. Dent. Assoc.* **1947**, *35*, 805–807. [CrossRef]

11. Liu, G.Y.; Nizet, V. Color me bad: Microbial pigments as virulence factors. *Trends Microbiol.* **2009**, *17*, 406–413. [CrossRef] [PubMed]

12. Haffajee, A.D.; Cugini, M.A.; Tanner, A.; Pollack, R.P.; Smith, C.; Kent, R.L., Jr.; Socransky, S.S. Subgingival microbiota in healthy, well-maintained elder and periodontitis subjects. *J. Clin. Periodontol.* **1998**, *25*, 346–353. [CrossRef] [PubMed]

13. Kononen, E.; Kanervo, A.; Takala, A.; Asikainen, S.; Jousimies-Somer, H. Establishment of oral anaerobes during the first year of life. *J. Dent. Res.* **1999**, *78*, 1634–1639. [CrossRef] [PubMed]

14. Willems, A.; Collins, M.D. Reclassification of *Oribaculum catoniae* (Moore and Moore 1994) as *Porphyromonas catoniae* comb. nov. and emendation of the genus *Porphyromonas*. *Int. J. Syst. Bacteriol.* **1995**, *45*, 578–581. [CrossRef]

15. Handley, P.S.; Correia, F.F.; Russell, K.; Rosan, B.; DiRienzo, J.M. Association of a novel high molecular weight, serine-rich protein (SrpA) with fibril-mediated adhesion of the oral biofilm bacterium *Streptococcus cristatus*. *Oral Microbiol. Immunol.* **2005**, *20*, 131–140. [CrossRef]

16. Rickard, A.H.; Gilbert, P.; High, N.J.; Kolenbrander, P.E.; Handley, P.S. Bacterial coaggregation: An integral process in the development of multi-species biofilms. *Trends Microbiol.* **2003**, *11*, 94–100. [CrossRef]

17. Kreth, J.; Zhang, Y.; Herzberg, M.C. Streptococcal antagonism in oral biofilms: *Streptococcus sanguinis* and *Streptococcus gordonii* interference with *Streptococcus mutans*. *J. Bacteriol.* **2008**, *190*, 4632–4640. [CrossRef]

18. Li, Y.; Zhang, Q.; Zhang, F.; Liu, R.; Liu, H.; Chen, F. Analysis of the Microbiota of Black Stain in the Primary Dentition. *PLoS ONE* **2015**, *10*, e0137030. [CrossRef]

19. Li, Y.; Zou, C.G.; Fu, Y.; Li, Y.; Zhou, Q.; Liu, B.; Zhang, Z.; Liu, J. Oral microbial community typing of caries and pigment in primary dentition. *BMC Genom.* **2016**, *17*, 558. [CrossRef]

20. Washio, J.; Shimada, Y.; Yamada, M.; Sakamaki, R.; Takahashi, N. Effects of pH and lactate on hydrogen sulfide production by oral *Veillonella* spp. *Appl. Environ. Microbiol.* **2014**, *80*, 4184–4188. [CrossRef]

21. Yoshiaki, N.; Ryoko, O.; Ryo, H.; Nobuhiro, H. Oral Microbiome of Children Living in an Isolated Area in Myanmar. *Int. J. Environ. Res.* **2020**, *17*, 4033.

22. Yoshiaki, N.; Erika, K.; Ayako, O.; Ryoko, O.; Mieko, S.; Yasuko, T.; Chieko, T.; Kazumune, A.; Hideki, D.; Tamotsu, S.; et al. Oral Microbiome in Four female Centenarians. *Appl. Sci.* **2020**, *10*, 5312.

23. Yoshiaki, N.; Erika, K.; Noboru, K.; Kaname, N.; Akihiro, Y.; Nobuhiro, H. The Oral Microbiome of Healthy Japanese People at the Age of 90. *Appl. Sci.* **2020**, *10*, 6450.

Publisher's Note: MDPI stays neutral with regard to jurisdictional claims in published maps and institutional affiliations.

Article

The Oral Microbiome of Healthy Japanese People at the Age of 90

Yoshiaki Nomura [1,*], Erika Kakuta [2], Noboru Kaneko [3], Kaname Nohno [3], Akihiro Yoshihara [4] and Nobuhiro Hanada [1]

1   Department of Translational Research, Tsurumi University School of Dental Medicine, Kanagawa 230-8501, Japan; hanada-n@tsurumi-u.ac.jp
2   Department of Oral bacteriology, Tsurumi University School of Dental Medicine, Kanagawa 230-8501, Japan; kakuta-erika@tsurumi-u.ac.jp
3   Division of Preventive Dentistry, Faculty of Dentistry and Graduate School of Medical and Dental Science, Niigata University, Niigata 951-8514, Japan; nkaneko@dent.niigata-u.ac.jp (N.K.); no2@dent.niigata-u.ac.jp (K.N.)
4   Division of Oral Science for Health Promotion, Faculty of Dentistry and Graduate School of Medical and Dental Science, Niigata University, Niigata 951-8514, Japan; akihiro@dent.niigata-u.ac.jp
*   Correspondence: nomura-y@tsurumi-u.ac.jp; Tel.: +81-45-580-8462

Received: 19 August 2020; Accepted: 15 September 2020; Published: 16 September 2020

Abstract: For a healthy oral cavity, maintaining a healthy microbiome is essential. However, data on healthy microbiomes are not sufficient. To determine the nature of the core microbiome, the oral-microbiome structure was analyzed using pyrosequencing data. Saliva samples were obtained from healthy 90-year-old participants who attended the 20-year follow-up Niigata cohort study. A total of 85 people participated in the health checkups. The study population consisted of 40 male and 45 female participants. Stimulated saliva samples were obtained by chewing paraffin wax for 5 min. The V3–V4 hypervariable regions of the 16S ribosomal RNA (rRNA) gene were amplified by PCR. Pyrosequencing was performed using MiSeq. Operational taxonomic units (OTUs) were assigned on the basis of a 97% identity search in the EzTaxon-e database. Using the threshold of 100% detection on the species level, 13 species were detected: *Streptococcus sinensis*, *Streptococcus pneumoniae*, *Streptococcus salivarius*, KV831974_s, *Streptococcus parasanguinis*, *Veillonella dispar*, *Granulicatella adiacens*, *Streptococcus*_uc, *Streptococcus peroris*, KE952139_s, *Veillonella parvula*, *Atopobium parvulum*, and AFQU_vs. These species represent potential candidates for the core make-up of the human microbiome.

Keywords: oral microbiome; pyrosequencing; core microbiome; elderly

1. Introduction

Microbiota inhabiting the human body have long been recognized as critical for a variety of human diseases and in maintaining human health [1–5]. The microbiome of humans has been extensively studied using accurate and efficient approaches involving high-throughput sequencing technologies and bioinformatics [6].

Among several microbiomes in the human body, the oral microbiome has been extensively studied. Specific bacterial taxa are responsible for oral infectious diseases, such as dental caries and periodontal diseases. Microbial diversity increases in parallel with the progression of periodontitis [7]. Several studies showed that there is a relationship between the human oral microbiome and certain systemic diseases, such as pancreatic cancer [8], Type 2 diabetes [9], pediatric Crohn's disease [10], heart disease [11], and low-weight preterm birth [12]. Certain oral bacterial species may contribute to carcinogenesis [13]. The role of the oral microbiome involves the mediation of inflammation related to changes in systemic health and disease [14].

In addition to these intensive studies related to disease, health-associated species in the oral cavity have been identified. One of the primary goals of the Human Microbiome Project [15] launched by the National Institutes of Health was to characterize the core microbiome. The concept of the core human oral microbiome involves comprehensive, minimal bacteria that regularly inhabit the human body. The human core microbiome is hypothesized to be important for development, health, and functioning. Some diseases, including autoimmune disorders [16], diabetes, and obesity [17,18], are caused by perturbation of the core gut microbiome. Therefore, a beneficial oral microbiome and its associated ecosystem functions may ensure host health and wellbeing [19–21].

The aims of this study were to determine if a healthy core microbiome is meaningful in the context of disease prevention and to investigate whether the microbiome of healthy older people may be a suitable representation of a healthy microbiome. Data on the oral microbiome of healthy older people may be useful for a comparison with the microbiome of subjects with specific diseases in an effort to determine if their etiology is related to an imbalance of the microbiome or colonization by specific bacterial taxa. These data may be useful for the development of a healthy core microbiome from a young age. Several studies have investigated the oral microbiome of older people. These studies were focused on the etiology of diseases or disease conditions [22–31]. Few studies investigating healthy older people are available [32–34].

The aim of this study was to investigate the oral microbiome of a community of healthy people at the age of 90 in an attempt to identify their core oral microbiome.

2. Materials and Methods

2.1. Study Design, Setting, and Participants

The subjects that participated in this study were sourced from the Niigata study, a community-based cohort study investigating the relationship between oral health and systemic health in older subjects aged 70 in 2008. Sampling and data collection procedures were described in previous reports in detail [35,36]. A 20-year follow-up study was conducted in 2018 through mass examination. Dental and medical examinations were carried out. Briefly, all of our cohorts did not present with any comorbidities.

2.2. Sampling and Microbial DNA Extraction

Saliva samples were collected by chewing paraffin wax for 5 min. Collected samples were kept on ice. Upon transporting them to the laboratory, samples were frozen at −20 °C until DNA extraction. Saliva samples were centrifuged at 3000 rpm for 10 min. The Maxwell 16 LEV Blood DNA Kit (Promega KK, Tokyo, Japan) was used for DNA extraction. The NanoDrop ND-2000 (Thermo Fisher Scientific KK, Tokyo, Japan) was used for the measurement of DNA concentration. DNA degradation was visually checked by electrophoresis on a 1% agarose gel using the Qubit dsDNA BR Assay Kit (Thermo Fisher Scientific KK, Tokyo, Japan). The inclusion criteria for DNA samples subjected to further analysis were as follows: concentration > 20 ng/μL, volume ≥ 20 μL, $A_{260/280}$ ≥ 1.8, and $A_{260/230}$ >1.5 [32,37].

2.3. Microbial-Community Analysis

A polymerase chain reaction (PCR) was carried out using primers specific to the V3–V4 region involving pyrosequencing tags of the 16S ribosomal RNA (rRNA) gene. The taxonomic classification of each read was assigned on the basis of a search of the EzBioCloud 16S database [38,39]. By applying the data from this database, a hierarchical taxonomic classification was obtained [32,37]. These analyses were carried out by Chun Lab (Seoul, Korea).

2.4. Bioinformatics Analysis

The relative abundance of the 16S rRNA gene for each operational taxonomic unit (OTU) was used to determine the absolute abundance of each OTU by multiplying the respective relative abundance by the total number of 16S rRNA gene copies. The Microbiome and Phyloseq packages in R software version 3.50 (Lucent Technologies, Murray Hill, NJ, USA) were used for analysis [40]. Heatmaps and core heatmaps [41] were used for visualization. Core line-plot and t-distributed stochastic neighbor embedding (t-SNE) [42] analyses were performed using the Microbiome Rtsne and Vegan packages.

2.5. Ethics Approval

All subjects who participated in this study were approved for the purpose of this study. Prior to saliva collection, they completed an informed-consent form. This study was approved by the Ethics Committee of the Tsurumi University School of Dental Medicine (Approval Number: 1332).

3. Results

3.1. Study Participants

The number of subjects participating in this study was 87. There were 41 men and 46 women who were all 90 years old. Adequate mount samples for pyrosequencing analysis were not obtained from one man and one woman.

3.2. Sequence Data

From the 85 subjects, 3,899,271 reads (minimum, 23,272; maximum, 108,597) passed quality control. From these reads, sequences were clustered into 24 phyla, 48 classes, 106 orders, 214 families, 529 genera, and 1216 species. The prevalence and abundances of all 1216 species are visualized using a heatmap in Figure S1 (Supplementary Materials).

The summary statistics of the alpha diversity indices are shown in Table S1 (Supplementary Materials). The rarefaction curve is presented in Figure S2 (Supplementary Materials).

3.3. Oral-Microbiome Structure

Figure 1 shows the relative abundance of detected bacteria. Data are presented separately on the (A) phylum and (B) genus levels. The Firmicutes phylum was most abundant, followed by Actinobacteria and Bacteroides. The abundance of these phyla constituted 92.6% of the sample. On the genus level, *Streptococcus* represented 44.5%, *Rothia* represented 15.2%, and *Veillonella* represented 9.0% of the sample. The composition bar plots for each sample on the phylum level are shown in Figure S3 (Supplementary Materials). The proportional ranges of these bacteria were 42.7% to 93.0% for Firmicutes, 4.7% to 39.8% for Actinobacteria, and <0.01% to 39.6% for Bacteroides. Taxon prevalence is shown in Figure 2. The prevalence of each species is plotted against their abundance. Highly prevalent phyla were Firmicutes and Actinobacteria. The core line plot is shown in Figure S4 (Supplementary Materials).

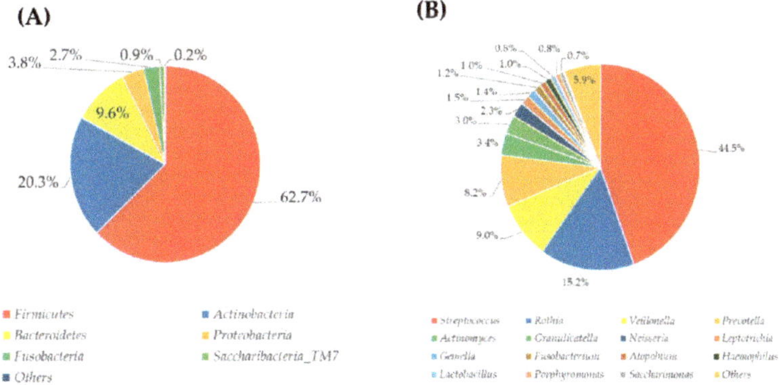

Figure 1. Microbiome structure in saliva: (**A**) phylum level; (**B**) genus level.

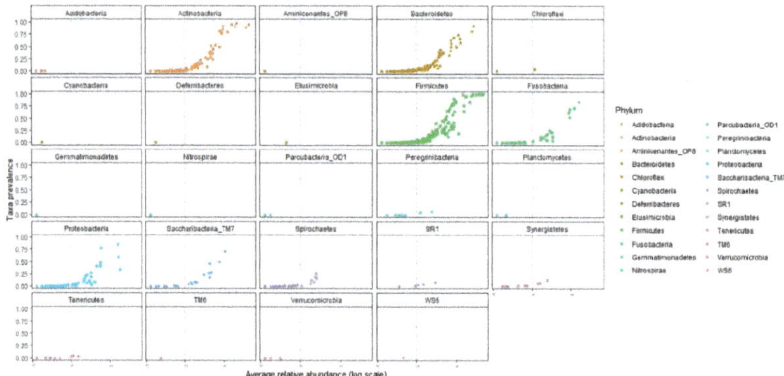

Figure 2. Overview of operational-taxonomic-unit (OTU) prevalence and taxonomic affiliations.

3.4. Candidates for Core Microbiome and Core-Microbiome Analysis

Table 1 shows the bacterial species detected in more than 90% of subjects. Thirteen species were detected in all subjects. Among these 13 species, eight species were from the *Streptococcus* genus. These species represent potential candidates for the core oral microbiome. These 13 species are visualized using a phylogenetic tree and core heatmap in Figure 3.

Table 1. Bacterial species detected in more than 90% of subjects.

	Taxonomy			Prevalence (*n*%)		Abundance (Average)
	Phylum	**Genus**	**Species**			
100%	Firmicutes	*Streptococcus*	*Streptococcus sinensis*	85/85	100%	10.14%
	Firmicutes	*Streptococcus*	*Streptococcus pneumoniae*	85/85	100%	9.61%
	Firmicutes	*Streptococcus*	*Streptococcus salivarius*	85/85	100%	8.75%
	Actinobacteria	*Rothia*	KV831974_s	85/85	100%	8.30%
	Firmicutes	*Streptococcus*	*Streptococcus parasanguinis*	85/85	100%	6.13%
	Firmicutes	*Veillonella*	*Veillonella dispar*	85/85	100%	4.31%
	Firmicutes	*Granulicatella*	*Granulicatella adiacens*	85/85	100%	2.90%
	Firmicutes	*Streptococcus*	*Streptococcus_uc*	85/85	100%	2.80%
	Firmicutes	*Streptococcus*	*Streptococcus peroris*	85/85	100%	2.67%
	Actinobacteria	*Actinomyces*	KE952139_s	85/85	100%	1.72%
	Firmicutes	*Veillonella*	*Veillonella parvula*	85/85	100%	1.15%
	Actinobacteria	*Atopobium*	*Atopobium parvulum*	85/85	100%	0.89%
	Firmicutes	*Streptococcus*	AFQU_s	85/85	100%	0.77%
99%–95%	Firmicutes	*Veillonella*	*Veillonella atypica*	84/85	98.8%	2.22%
	Actinobacteria	*Actinomyces*	*Actinomyces_uc*	84/85	98.8%	0.30%
	Firmicutes	*Veillonella*	*Veillonella_uc*	84/85	98.8%	0.25%
	Actinobacteria	*Rothia*	*Rothia mucilaginosa*	83/85	97.6%	5.43%
	Firmicutes	*Gemella*	*Gemella haemolysans*	83/85	97.6%	1.23%
	Actinobacteria	*Rothia*	*Rothia_uc*	83/85	97.6%	0.28%
	Proteobacteria	*Campylobacter*	*Campylobacter concisus*	83/85	97.6%	0.10%
	Bacteroides	*Prevotella*	*Prevotella melaninogenica*	82/85	96.5%	2.92%
	Firmicutes	*Streptococcus*	*Streptococcus gordonii*	82/85	96.5%	1.78%
	Firmicutes	*Moryella*	*Stomatobaculum longum*	82/85	96.5%	0.31%
	Actinobacteria	*Actinomyces*	JVLH_s	82/85	96.5%	0.18%
	Actinobacteria	*Rothia*	*Rothia dentocariosa*	81/85	95.3%	1.36%
	Firmicutes	*Bulleidia*	*Solobacterium moorei*	81/85	95.3%	0.20%
>90%	Fusobacteria	*Fusobacterium*	*Fusobacterium nucleatum*	80/85	94.1%	0.87%
	Firmicutes	*Megasphaera*	*Megasphaera micronuciformis*	80/85	94.1%	0.30%
	Actinobacteria	*Actinomyces*	*Actinomyces odontolyticus*	80/85	94.1%	0.24%
	Firmicutes	*Lachnoanaerobaculum*	*Lachnoanaerobaculum saburreum*	80/85	94.1%	0.17%
	Bacteroides	*Prevotella*	*Prevotella histicola*	79/85	92.9%	2.39%
	Proteobacteria	*Haemophilus*	*Haemophilus parainfluenzae*	79/85	92.9%	0.88%
	Bacteroides	*Prevotella*	*Prevotella_uc*	79/85	92.9%	0.42%
	Firmicutes	*Streptococcus*	*Streptococcus sanguinis*	78/85	91.8%	0.92%
	Actinobacteria	*Actinomyces*	*Actinomyces graevenitzii*	78/85	91.8%	0.52%
	Bacteroides	*Prevotella*	*Prevotella salivae*	77/85	90.6%	0.25%
	Firmicutes	*Oribacterium*	*Oribacterium asaccharolyticum*	77/85	90.6%	0.24%

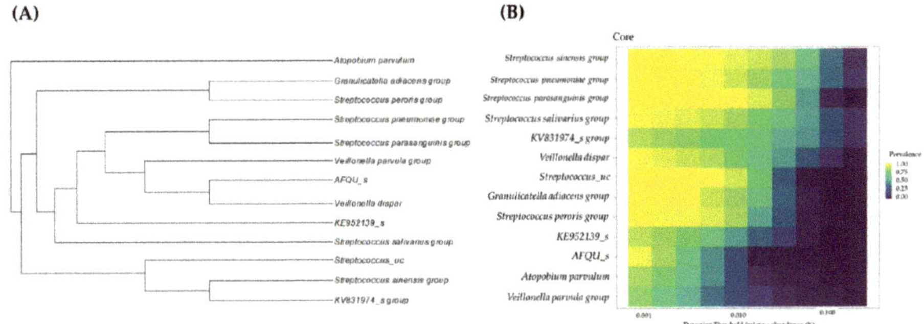

Figure 3. Core visualization: (**A**) phylogenetic tree and (**B**) core heatmap of 13 species detected in all subjects.

According to Figure 3, eight species of *Streptococcus* were not adjacent. *Streptococcus sinensis* and *Streptococcus pneumoniae* were highly prevalent and abundant species.

3.5. Ordination Analysis

The t-SNE plot allows embedding high-dimensional data into a two-dimensional map on the basis of each data point. Each species was categorized according to its prevalence (Figure 4A) or genus (Figure 4B). Through this graphical observation, highly prevalent species were conglomerated, and low-prevalence species were separated into distinct groups. Species were not separated on the basis of their genus. These results indicate that, on the species level, oral bacteria formed groups, and their coexistence was not regulated by their genus.

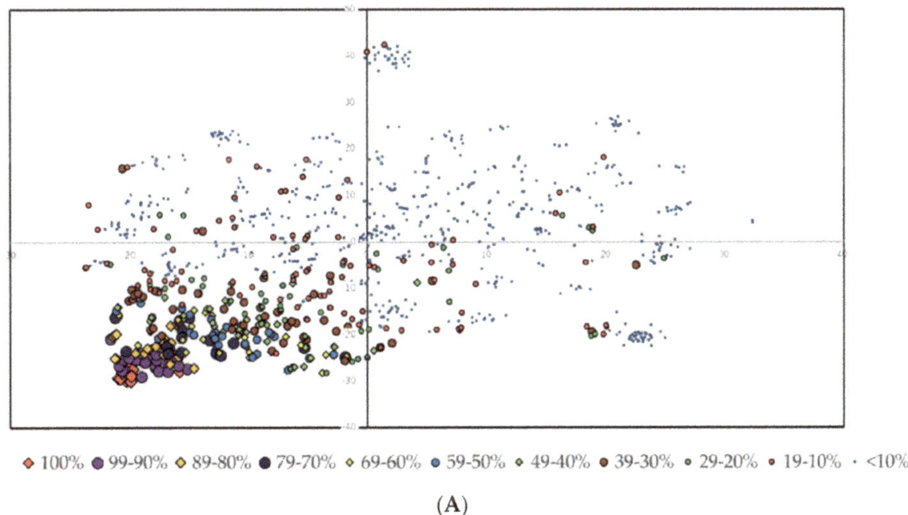

(**A**)

Figure 4. *Cont.*

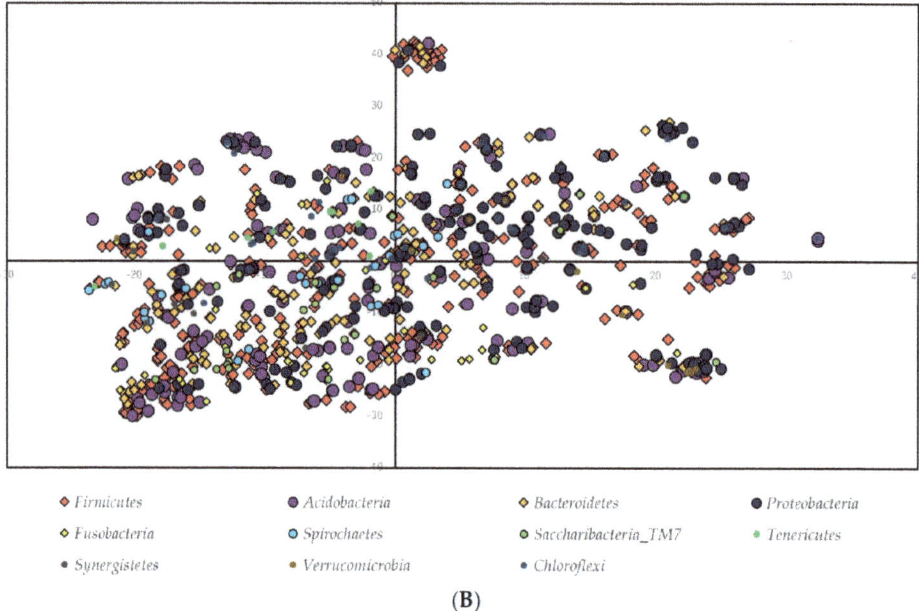

Figure 4. Ordination analysis as a function of t-distributed stochastic neighbor embedding (t-SNE) plot categorized by (**A**) prevalence and (**B**) genus.

4. Discussion

In this study, we investigated 85 healthy people at the age of 90, focusing on analyzing their core oral microbiome. On the basis of the pyrosequencing analysis, 389,927 valid reads were obtained, and 1216 species were detected. Thirteen species were detected in all 85 subjects. Among them, seven belonged to the *Streptococcus* genus. These species represent potential candidates for the core human oral microbiome.

In this study, the most abundant phylum was Firmicutes, whereas *Streptococcus* constituted 45% of the genera present. The microbiome structure on the phylum level was categorized as follows: *Firmicutes* (62.7%), *Bacteroides* (9.6%), *Proteobacteria* (3.8%), *Fusobacteria* (2.7%), and *Actinobacteria* (20.3%). A previous study investigating subjects aged 60 years and older in China showed that the abundant phyla were *Firmicutes* (29.6%), *Bacteroides* (22.4%), Proteobacteria (20.4%), Fusobacteria (16.2%), and Actinobacteria (7.6%) [43]. Another study showed that the core oral microbiome predominantly comprises Firmicutes, followed by Proteobacteria and Bacteroides [44]. *Firmicutes, Proteobacteria, Bacteroides, Fusobacteria,* and *Actinobacteria* constitute more than 98% of the oral microbiome [45]. The oral-microbiome structure may change as a function of growth [46], food [47,48], oral diseases [49–51], or infection [51,52]. These results are consistent with the results of this study.

Few reports presented the core microbiome on the genus or species level. On the genus level, *Neisseria* (12.5%), *Leptotrichia* (11.1%), *Streptococcus* (10.7%), *Prevotella* (7.0%), *Veillonella* (6.9%), *Fusobacterium* (5.4%), *Capnocytophaga* (4.2%), *Prevotella* (4.1%), *Corynebacterium* (2.6%), *Saccharibacteria* (2.6%), *Actinomyces* (2.6%), *Haemophilus* (2.3%), and *Porphyromonas* (2.2%) were most prevalent [43]. Abundant genera according to another study were *Streptococcus* (26.1%), *Veillonella* (21.9%), *Neisseria* (16.9%), *Haemophilus* (10.7%), *Actinomycetes* (2.6%), *Rothia* (3.1%), and *Oribacterium* (1.7%) [53]. On the species level, a study carried out in Japan identified *Streptococcus salivarius* (9.5%), *Prevotella melaninogenica* (9.2%), *Rothia mucilaginosa* (8.8%), *Veillonella atypica* (6.0%), and *Neisseria flavescens* (5.8%) as species exhibiting > 5% abundance on the tongue surface [34]. When compared with our results, shown in Table 1, four of these species were detected in all our subjects, except for

Neisseria flavescens. *Neisseria* sp. is often detected in samples obtained from the dorsal surface of the tongue [54]. On the phylum level, the core microbiome observed in our study coincided with that in the literature. However, on the genus or species level, predominant bacteria varied across studies.

There are environmental and cultural differences, such as food consumption, that affect microbiome structure [49,55–57]; in our study, the proportions of Firmicutes and *Streptococcus* were higher than those found in other studies.

Streptococcus spp. are abundant in human milk, and they play an important role in the establishment of the oral-microbiome structure for breastfed infants [58,59]. Some *Streptococcus* spp. act as probiotic bacteria [60,61]. In contrast to these beneficial effects of *Streptococcus* for the human body, pathogenic *Streptococcus sinensis*, which is responsible for bacteremia [62] and endocarditis [63], and *Streptococcus pneumoniae* [64] were detected in all samples.

Two species of *Veillonella*, which are also classified as Firmicutes, were detected in all samples. *Veillonella* is known to be prevalently detected at various sites in the oral cavity, such as dental plaque, saliva [65], and oral mucosa [66]. *Veillonella parvula* is associated with the development of dental caries [67], endodontic infections [68], and periodontitis [69]. The most abundant species in our study was *Streptococcus sinensis*, followed by *Streptococcus pneumoniae*.

In this study, *Fusobacterium nucleatum* was detected in 94.1% of subjects. A previous study showed that Fusobacteria represent a predominant taxon in the oral microbiome [69]. Fusobacteria mediate the coaggregation of nonaggregating microbiota, and they are a structural element of plaque in both healthy and disease conditions [70]. They may contribute to the diversity of the oral microbiome.

A pioneering study of the oral microbiome using pyrosequencing suggested that the concept of a healthy core microbiome was supported by abundant oral taxa found in the oral cavity of healthy individuals [66,71]. Phylogenetic trees of 118 of the most predominant taxa identified at several sampling sites in the oral cavity were presented [71]. Among the 13 species detected in all subjects, five species were not included in this phylogenetic tree. These 13 species included pathogenic bacteria leading to human diseases. These bacteria were also detected in healthy older persons; however, they are not proposed as candidates for a healthy core oral microbiome.

It has been suggested that periodontal disease can be a major risk factor for some systemic diseases [72–77]. Recent advances in research on oral and general health have shown that there are protective host factors for periodontal-related systemic diseases [78–80]. The limitation of this study is its cross-sectional study. Further study is necessary to investigate the oral microbiome that can be the risk for mortality in combination with these host factors for older people.

In this study, we aimed to identify the core oral microbiome in healthy older people. However, on the basis of prevalence and abundance, pathogenic bacteria were also included. The human oral microbiome plays a crucial role in diseases and human health. Simple descriptive analysis as a function of prevalence and abundance may not be enough to define a healthy core microbiome. The effect of bacteria should be considered when defining a healthy human oral microbiome.

Supplementary Materials: The following are available online at http://www.mdpi.com/2076-3417/10/18/6450/s1. Figure S1. Heatmap of 1216 species detected in this study; Figure S2. Rarefaction curves of the 85 subjects; Figure S3. Composition bar plots for each subject at the phylum level; Figure S4. Core line plot; Table S1. Alpha diversity indices.

Author Contributions: Conceptualization, Y.N., A.Y., and N.H.; methodology, Y.N. and A.Y.; formal analysis, Y.N.; investigation, Y.N., E.K., and M.O.; resources, N.K.; writing—original-draft preparation, Y.N.; writing—review and editing, Y.N.; visualization, Y.N.; project administration, K.N.; funding acquisition, Y.N. and N.H. All authors read and agreed to the published version of the manuscript.

Funding: This study was supported by the JSPS KAKENHI (grant numbers 17K12030 and 20K10303) and the SECOM Science and Technology Foundation. The funders played no role in the design of the study, in data collection, in the analysis and interpretation of the results, or in the writing of the manuscript.

Conflicts of Interest: The authors declare no conflict of interest.

References

1. Lynch, S.V.; Pedersen, O. The Human Intestinal Microbiome in Health and Disease. *N. Engl. J. Med.* **2016**, *375*, 2369–2379. [CrossRef] [PubMed]
2. Silbergeld, E.K. The Microbiome. *Toxicol. Pathol.* **2017**, *45*, 190–194. [CrossRef] [PubMed]
3. Tuddenham, S.; Sears, C.L. The intestinal microbiome and health. *Curr. Opin. Infect. Dis.* **2015**, *28*, 464–470. [CrossRef] [PubMed]
4. Singh, R.K.; Chang, H.W.; Yan, D.; Lee, K.M.; Ucmak, D.; Wong, K.; Abrouk, M.; Farahnik, B.; Nakamura, M.; Zhu, T.H.; et al. Influence of diet on the gut microbiome and implications for human health. *J. Transl. Med.* **2017**, *15*, 73. [CrossRef]
5. Rinninella, E.; Raoul, P.; Cintoni, M.; Franceschi, F.; Miggiano, G.A.D.; Gasbarrini, A.; Mele, M.C. What is the Healthy Gut Microbiota Composition? A Changing Ecosystem across Age, Environment, Diet, and Diseases. *Microorganisms* **2019**, *7*, 14. [CrossRef]
6. Schwartz, M.H.; Wang, H.; Pan, J.N.; Clark, W.C.; Cui, S.; Eckwahl, M.J.; Pan, D.W.; Parisien, M.; Owens, S.M.; Cheng, B.L.; et al. Microbiome characterization by high-throughput transfer RNA sequencing and modification analysis. *Nat. Commun.* **2018**, *9*, 5353. [CrossRef]
7. Papapanou, P.N.; Park, H.; Cheng, B.; Kokaras, A.; Paster, B.; Burkett, S.; Watson, C.W.; Annavajhala, M.K.; Uhlemann, A.C.; Noble, J.M. Subgingival microbiome and clinical periodontal status in an elderly cohort: The WHICAP ancillary study of oral health. *J. Periodontol.* **2020**. [CrossRef]
8. Farrell, J.J.; Zhang, L.; Zhou, H.; Chia, D.; Elashoff, D.; Akin, D.; Paster, B.J.; Joshipura, K.; Wong, D.T. Variations of oral microbiota are associated with pancreatic diseases including pancreatic cancer. *Gut* **2012**, *61*, 582–588. [CrossRef]
9. Demmer, R.T.; Jacobs, D.R., Jr.; Singh, R.; Zuk, A.; Rosenbaum, M.; Papapanou, P.N.; Desvarieux, M. Periodontal Bacteria and Prediabetes Prevalence in ORIGINS: The Oral Infections, Glucose Intolerance, and Insulin Resistance Study. *J. Dent. Res.* **2015**, *94*, 201S–211S. [CrossRef]
10. Docktor, M.J.; Paster, B.J.; Abramowicz, S.; Ingram, J.; Wang, Y.E.; Correll, M.; Jiang, H.; Cotton, S.L.; Kokaras, A.S.; Bousvaros, A. Alterations in diversity of the oral microbiome in pediatric inflammatory bowel disease. *Inflamm. Bowel Dis.* **2012**, *18*, 935–942. [CrossRef]
11. Liu, X.R.; Xu, Q.; Xiao, J.; Deng, Y.M.; Tang, Z.H.; Tang, Y.L.; Liu, L.S. Role of oral microbiota in atherosclerosis. *Clin. Chim. Acta* **2020**, *506*, 191–195. [CrossRef] [PubMed]
12. Cobb, C.M.; Kelly, P.J.; Williams, K.B.; Babbar, S.; Angolkar, M.; Derman, R.J. The oral microbiome and adverse pregnancy outcomes. *Int. J. Womens Health* **2017**, *9*, 551–559. [CrossRef] [PubMed]
13. Teles, F.R.F.; Alawi, F.; Castilho, R.M.; Wang, Y. Association or Causation? Exploring the Oral Microbiome and Cancer Links. *J. Dent. Res.* **2020**. [CrossRef] [PubMed]
14. Kleinstein, S.E.; Nelson, K.E.; Freire, M. Inflammatory Networks Linking Oral Microbiome with Systemic Health and Disease. *J. Dent. Res.* **2020**, *99*, 1131–1139. [CrossRef] [PubMed]
15. Integrative HMP (iHMP) Research Network Consortium. The Integrative Human Microbiome Project. *Nature* **2019**, *569*, 641–648. [CrossRef] [PubMed]
16. Severance, E.G.; Yolken, R.H.; Eaton, W.W. Autoimmune diseases, gastrointestinal disorders and the microbiome in schizophrenia: More than a gut feeling. *Schizophr. Res.* **2016**, *176*, 23–35. [CrossRef] [PubMed]
17. Barengolts, E.; Green, S.J.; Chlipala, G.E.; Layden, B.T.; Eisenberg, Y.; Priyadarshini, M.; Dugas, L.R. Predictors of Obesity among Gut Microbiota Biomarkers in African American Men with and without Diabetes. *Microorganisms* **2019**, *7*, 320. [CrossRef]
18. Zhang, F.; Wang, M.; Yang, J.; Xu, Q.; Liang, C.; Chen, B.; Zhang, J.; Yang, Y.; Wang, H.; Shang, Y.; et al. Response of gut microbiota in type 2 diabetes to hypoglycemic agents. *Endocrine* **2019**, *66*, 485–493. [CrossRef]
19. Ulloa, C.P.; van der Veen, M.H.; Krom, B.P. Review: Modulation of the oral microbiome by the host to promote ecological balance. *Odontology* **2019**, *107*, 437–448. [CrossRef]
20. Marsh, P.D. In Sickness and in Health-What Does the Oral Microbiome Mean to Us? An Ecological Perspective. *Adv. Dent. Res.* **2018**, *29*, 60–65. [CrossRef]
21. Rosier, B.T.; Marsh, P.D.; Mira, A. Resilience of the Oral Microbiota in Health: Mechanisms That Prevent Dysbiosis. *J. Dent. Res.* **2018**, *97*, 371–380. [CrossRef] [PubMed]
22. Tam, J.; Hoffmann, T.; Fischer, S.; Bornstein, S.; Gräßler, J.; Noack, B. Obesity alters composition and diversity of the oral microbiota in patients with type 2 diabetes mellitus independently of glycemic control. *PLoS ONE* **2018**, *13*, e0204724. [CrossRef]

23. Abe, K.; Takahashi, A.; Fujita, M.; Imaizumi, H.; Hayashi, M.; Okai, K.; Ohira, H. Dysbiosis of oral microbiota and its association with salivary immunological biomarkers in autoimmune liver disease. *PLoS ONE* **2018**, *13*, e0198757. [CrossRef] [PubMed]

24. Gaetti-Jardim, E., Jr.; Jardim, E.C.; Schweitzer, C.M.; da Silva, J.C.; Oliveira, M.M.; Masocatto, D.C.; dos Santos, C.M. Supragingival and subgingival microbiota from patients with poor oral hygiene submitted to radiotherapy for head and neck cancer treatment. *Arch. Oral Biol.* **2018**, *90*, 45–52. [CrossRef] [PubMed]

25. Si, J.; Lee, C.; Ko, G. Oral Microbiota: Microbial Biomarkers of Metabolic Syndrome Independent of Host Genetic Factors. *Front. Cell. Infect. Microbiol.* **2017**, *7*, 516. [CrossRef] [PubMed]

26. Ogawa, T.; Ogawa, H.M.; Ikebe, K.; Notomi, Y.; Iwamoto, Y.; Shirobayashi, I.; Hata, S.; Kibi, M.; Masayasu, S.; Sasaki, S.; et al. Characterizations of oral microbiota in elderly nursing home residents with diabetes. *J. Oral Sci.* **2017**, *59*, 549–555. [CrossRef]

27. Flemer, B.; Warren, R.D.; Barrett, M.P.; Cisek, K.; Das, A.; Jeffery, I.B.; Hurley, E.; O'Riordain, M.; Shanahan, F.; O'Toole, P.W. The oral microbiota in colorectal cancer is distinctive and predictive. *Gut* **2018**, *67*, 1454–1463. [CrossRef]

28. Olson, S.H.; Satagopan, J.; Xu, Y.; Ling, L.; Leong, S.; Orlow, I.; Saldia, A.; Li, P.; Nunes, P.; Madonia, V.; et al. The oral microbiota in patients with pancreatic cancer, patients with IPMNs, and controls: A pilot study. *Cancer Causes Control* **2017**, *28*, 959–969. [CrossRef]

29. Zhu, X.X.; Yang, X.J.; Chao, Y.L.; Zheng, H.M.; Sheng, H.F.; Liu, H.Y.; He, Y.; Zhou, H.W. The Potential Effect of Oral Microbiota in the Prediction of Mucositis during Radiotherapy for Nasopharyngeal Carcinoma. *EBioMedicine* **2017**, *18*, 23–31. [CrossRef]

30. Chen, X.; Winckler, B.; Lu, M.; Cheng, H.; Yuan, Z.; Yang, Y.; Jin, L.; Ye, W. Oral Microbiota and Risk for Esophageal Squamous Cell Carcinoma in a High-Risk Area of China. *PLoS ONE* **2015**, *10*, e0143603. [CrossRef]

31. Gaiser, R.A.; Halimi, A.; Alkharaan, H.; Lu, L.; Davanian, H.; Healy, K.; Hugerth, L.W.; Ateeb, Z.; Valente, R.; Fernandez Moro, F.C.; et al. Enrichment of oral microbiota in early cystic precursors to invasive pancreatic cancer. *Gut* **2019**, *68*, 2186–2194. [CrossRef] [PubMed]

32. Nomura, Y.; Otsuka, R.; Hasegawa, R.; Hanada, N. Oral Microbiome of Children Living in an Isolated Area in Myanmar. *Int. J. Environ. Res. Public Health* **2020**, *17*, 4033. [CrossRef] [PubMed]

33. Iwauchi, M.; Horigome, A.; Ishikawa, K.; Mikuni, A.; Nakano, M.; Xiao, J.Z.; Odamaki, T.; Hironaka, S. Relationship between oral and gut microbiota in elderly people. *Immun. Inflamm. Dis.* **2019**, *7*, 229–236. [CrossRef]

34. Asakawa, M.; Takeshita, T.; Furuta, M.; Kageyama, S.; Takeuchi, K.; Hata, J.; Ninomiya, T.; Yamashita, Y. Tongue Microbiota and Oral Health Status in Community-Dwelling Elderly Adults. *mSphere* **2018**, *3*, e00332-18. [CrossRef]

35. Hirotomi, T.; Yoshihara, A.; Ogawa, H.; Ito, K.; Igarashi, A.; Miyazaki, H. A preliminary study on the relationship between stimulated saliva and periodontal conditions in community-dwelling elderly people. *J. Dent.* **2006**, *34*, 692–698. [CrossRef] [PubMed]

36. Yamaga, T.; Ogawa, H.; Miyazaki, H. Influence of occlusal deterioration considering prosthetics on subsequent all-cause mortality in a Japanese elderly independent population. *Gerodontology* **2019**, *36*, 163–170. [CrossRef] [PubMed]

37. Nomura, Y.; Kakuta, E.; Okada, A.; Otsuka, R.; Shimada, M.; Tomizawa, Y.; Taguchi, C.; Arikawa, K.; Daikoku, H.; Sato, T.; et al. Oral microbiome in four female centenarians. *Appl. Sci.* **2020**, *10*, 5312. [CrossRef]

38. Kim, O.S.; Cho, Y.J.; Lee, K.; Yoon, S.H.; Kim, M.; Na, H.; Park, S.C.; Jeon, Y.S.; Lee, J.H.; Yi, H.; et al. Introducing EzTaxon-e: A prokaryotic 16S rRNA gene sequence database with phylotypes that represent uncultured species. *Int. J. Syst. Evol. Microbiol.* **2012**, *62*, 716–721. [CrossRef]

39. Yoon, S.H.; Ha, S.M.; Kwon, S.; Lim, J.; Kim, Y.; Seo, H.; Chun, J. Introducing EzBioCloud: A taxonomically united database of 16S rRNA and whole genome assemblies. *Int. J. Syst. Evol. Microbiol.* **2017**, *67*, 1613–1617. [CrossRef]

40. Lahti, L.; Shetty, S. Introduction to the Microbiome R Package. 2020. Available online: https://microbiome.github.io/tutorials/ (accessed on 31 August 2020).

41. Shetty, S.A.; Hugenholtz, F.; Lahti, L.; Smidt, H.; de Vos, W.M. Intestinal microbiome landscaping: Insight in community assemblage and implications for microbial modulation strategies. *FEMS Microbiol. Rev.* **2017**, *41*, 182–199. [CrossRef]

42. Bagwell, C.B. High-Dimensional Modeling for Cytometry: Building Rock Solid Models Using GemStone™ and Verity Cen-se'™ High-Definition t-SNE Mapping. *Methods Mol. Biol.* **2018**, *1678*, 11–36. [CrossRef]

43. Jiang, Q.; Liu, J.; Chen, L.; Gan, N.; Yang, D. The Oral Microbiome in the Elderly with Dental Caries and Health. *Front. Cell. Infect. Microbiol.* **2019**, *8*, 442. [CrossRef] [PubMed]

44. Balle, C.; Esra, R.; Havyarimana, E.; Jaumdally, S.Z.; Lennard, K.; Konstantinus, I.N.; Barnabas, S.L.; Happel, A.U.; Gill, K.; Pidwell, T.; et al. Relationship between the Oral and Vaginal Microbiota of South African Adolescents with High Prevalence of Bacterial Vaginosis. *Microorganisms* **2020**, *8*, 1004. [CrossRef]

45. Matsha, T.E.; Prince, Y.; Davids, S.; Chikte, U.; Erasmus, R.T.; Kengne, A.P.; Davison, G.M. Oral Microbiome Signatures in Diabetes Mellitus and Periodontal Disease. *J. Dent. Res.* **2020**, *99*, 658–665. [CrossRef]

46. Burcham, Z.M.; Garneau, N.L.; Comstock, S.S.; Tucke, R.M.; Knight, R.; Metcalf, J.L. Patterns of Oral Microbiota Diversity in Adults and Children: A Crowdsourced Population Study. *Sci. Rep.* **2020**, *10*, 2133. [CrossRef] [PubMed]

47. Nakano, M.; Wakabayashi, H.; Sugahara, H.; Odamaki, T.; Yamauchi, K.; Abe, F.; Xiao, J.Z.; Murakami, K.; Ishikawa, K.; Hironaka, S. Effects of lactoferrin and lactoperoxidase-containing food on the oral microbiota of older individuals. *Microbiol. Immunol.* **2017**, *61*, 416–426. [CrossRef]

48. Anderson, A.C.; Rothballer, M.; Altenburger, M.J.; Woelber, J.P.; Karygianni, L.; Vach, K.; Hellwig, E.; Al-Ahmad, A. Long-term fluctuation of oral biofilm microbiota following different dietary phases. *Appl. Environ. Microbiol.* **2020**. [CrossRef]

49. Sampaio-Maia, B.; Monteiro-Silva, F. Acquisition and maturation of oral microbiome throughout childhood: An update. *Dent. Res. J.* **2014**, *11*, 291–301.

50. Anderson, A.C.; Rothballer, M.; Altenburger, M.J.; Woelber, J.P.; Karygianni, L.; Lagkouvardos, I.; Hellwig, E.; Al-Ahmad, A. In-vivo shift of the microbiota in oral biofilm in response to frequent sucrose consumption. *Sci. Rep.* **2018**, *8*, 14202. [CrossRef]

51. Yang, L.; Dunlap, D.G.; Qin, S.; Fitch, A.; Li, K.; Koch, C.D.; Nouraie, M.; DeSensi, R.; Ho, K.S.; Martinson, J.J.; et al. Alterations in Oral Microbiota in HIV Are Related to Decreased Pulmonary Function. *Am. J. Respir. Crit. Care Med.* **2020**, *201*, 445–457. [CrossRef]

52. Starr, J.R.; Huang, Y.; Lee, K.H.; Murphy, C.M.; Moscicki, A.B.; Shiboski, C.H.; Ryder, M.I.; Yao, T.J.; Faller, L.L.; Dyke, V.R.B.; et al. Oral microbiota in youth with perinatally acquired HIV infection. *Microbiome* **2018**, *6*, 100. [CrossRef] [PubMed]

53. Anbalagan, R.; Srikanth, P.; Mani, M.; Barani, R.; Seshadri, K.G.; Janarthanan, R. Next generation sequencing of oral microbiota in Type 2 diabetes mellitus prior to and after neem stick usage and correlation with serum monocyte chemoattractant-1. *Diabetes Res. Clin. Pract.* **2017**, *130*, 204–210. [CrossRef] [PubMed]

54. Querido, N.B.; de Araujo, W.C. Selective isolation of Neisseria sicca from the human oral cavity on eosin methylene blue agar. *Appl. Environ. Microbiol.* **1976**, *4*, 612–614. [CrossRef] [PubMed]

55. Cephas, K.D.; Kim, J.; Mathai, R.A.; Barry, K.A.; Dowd, S.E.; Meline, B.S.; Swanson, K.S. Comparative analysis of salivary bacterial microbiome diversity in edentulous infants and their mothers or primary care givers using pyrosequencing. *PLoS ONE* **2011**, *6*, e23503. [CrossRef]

56. Shi, W.; Qin, M.; Chen, F.; Xia, B. Supragingival Microbial Profiles of Permanent and Deciduous Teeth in Children with Mixed Dentition. *PLoS ONE* **2016**, *11*, e0146938. [CrossRef]

57. Lassalle, F.; Spagnoletti, M.; Fumagalli, M.; Shaw, L.; Dyble, M.; Walker, C.; Thomas, M.G.; Migliano, A.B.; Balloux, F. Oral microbiomes from hunter-gatherers and traditional farmers reveal shifts in commensal balance and pathogen load linked to diet. *Mol. Ecol.* **2018**, *27*, 182–195. [CrossRef]

58. Ruiz, L.; Bacigalupe, R.; García-Carral, C.; Boix-Amoros, A.; Argüello, H.; Silva, C.B.; de los Angeles Checa, M.; Mira, A.; Rodríguez, J. Microbiota of human precolostrum and its potential role as a source of bacteria to the infant mouth. *Sci. Rep.* **2019**, *9*, 8435. [CrossRef]

59. Biagi, E.; Aceti, A.; Quercia, S.; Beghetti, I.; Rampelli, S.; Turroni, S.; Soverini, M.; Zambrini, A.V.; Faldella, G.; Candela, M.; et al. Microbial Community Dynamics in Mother's Milk and Infant's Mouth and Gut in Moderately Preterm Infants. *Front. Microbiol.* **2018**, *22*, 2512. [CrossRef]

60. Bidossi, A.; de Grandi, R.; Toscano, M.; Bottagisio, M.; de Vecchi, E.; Gelardi, M.; Drago, L. Probiotics *Streptococcus salivarius* 24SMB and *Streptococcus oralis* 89a interfere with biofilm formation of pathogens of the upper respiratory tract. *BMC Infect. Dis.* **2018**, *18*, 653. [CrossRef]

61. Humphreys, G.J.; McBain, A.J. Antagonistic effects of *Streptococcus* and *Lactobacillus* probiotics in pharyngeal biofilms. *Lett. Appl. Microbiol.* **2019**, *68*, 303–312. [CrossRef]

62. Woo, P.C.; Teng, J.L.; Leung, K.W.; Lau, S.K.; Tse, H.; Wong, B.H.; Yuen, K.Y. *Streptococcus sinensis* may react with Lancefield group F antiserum. *J. Med. Microbiol.* **2004**, *53*, 1083–1088. [CrossRef] [PubMed]

63. Faibis, F.; Mihaila, L.; Perna, S.; Lefort, J.F.; Demachy, M.C.; le Flèche-Matéos, A.; Bouvet, A. Streptococcus sinensis: An emerging agent of infective endocarditis. *J. Med. Microbiol.* **2008**, *57*, 528–531. [CrossRef] [PubMed]

64. Leonard, A.; Lalk, M. Infection and metabolism—Streptococcus pneumoniae metabolism facing the host environment. *Cytokine* **2018**, *112*, 75–86. [CrossRef] [PubMed]

65. Valm, A.M.; Welch, J.L.; Rieken, C.W.; Hasegawa, Y.; Sogin, M.L.; Oldenbourg, R.; Dewhirst, F.E.; Borisy, G.G. Systems-level analysis of microbial community organization through combinatorial labeling and spectral imaging. *Proc. Natl. Acad. Sci. USA* **2011**, *108*, 4152–4157. [CrossRef] [PubMed]

66. Zaura, E.; Keijser, B.J.; Huse, S.M.; Crielaard, W. Defining the healthy "core microbiome" of oral microbial communities. *BMC Microbiol.* **2009**, *9*, 259. [CrossRef]

67. Kanasi, E.; Dewhirst, F.E.; Chalmers, N.I.; Kent, R., Jr.; Moore, A.; Hughes, C.V.; Pradhan, N.; Loo, C.Y.; Tanner, A.C.R. Clonal analysis of the microbiota of severe early childhood caries. *Caries Res.* **2010**, *44*, 485–497. [CrossRef]

68. Khemaleelaku, S.; Baumgartner, J.C.; Pruksakorn, S. Identification of bacteria in acute endodontic infections and their antimicrobial susceptibility. *Oral Surg. Oral Med. Oral Pathol. Oral Radiol. Endod.* **2002**, *94*, 746–755. [CrossRef]

69. Mashima, I.; Kamaguchi, A.; Nakazawa, F. The distribution and frequency of oral *veillonella* spp. in the tongue biofilm of healthy young adults. *Curr. Microbiol.* **2011**, *63*, 403–407. [CrossRef]

70. Kolenbrander, P.E. Oral microbial communities: Biofilms, interactions, and genetic systems. *Annu. Rev. Microbiol.* **2000**, *54*, 413–437. [CrossRef]

71. Aas, J.A.; Paster, B.J.; Stokes, L.N.; Olsen, I.; Dewhirst, F.E. Defining the normal bacterial flora of the oral cavity. *J. Clin. Microbiol.* **2005**, *43*, 5721–5732. [CrossRef]

72. Jensen, E.; Allen, G.; Bednarz, J.; Couper, J.; Peña, A. Periodontal risk markers in children and adolescents with type 1 diabetes: A systematic review and meta-analysis. *Diabetes Metab. Res. Rev.* **2020**, e3368. [CrossRef]

73. Wang, J.; Yang, X.; Zou, X.; Zhang, Y.; Wang, J.; Wang, Y. Relationship between periodontal disease and lung cancer: A systematic review and meta-analysis. *J. Periodontal. Res.* **2020**. [CrossRef] [PubMed]

74. Nadim, R.; Tang, J.; Dilmohamed, A.; Yuan, S.; Wu, C.; Bakre, A.T.; Partridge, M.; Ni, J.; Copeland, J.R.; Anstey, K.J.; et al. Influence of periodontal disease on risk of dementia: A systematic literature review and a meta-analysis. *Eur. J. Epidemiol.* **2020**. [CrossRef]

75. Priyamvara, A.; Dey, A.K.; Bandyopadhyay, D.; Katikineni, V.; Zaghlol, R.; Basyal, B.; Barssoum, K.; Amarin, R.; Bhatt, D.L.; Lavie, C.J. Periodontal Inflammation and the Risk of Cardiovascular Disease. *Curr. Atheroscler. Rep.* **2020**, *22*, 28. [CrossRef]

76. Orlandi, M.; Graziani, F.; D'Aiuto, F. Periodontal therapy and cardiovascular risk. *Periodontol. 2000* **2020**, *83*, 107–124. [CrossRef]

77. Jepsen, S.; Suvan, J.; Deschner, J. The association of periodontal diseases with metabolic syndrome and obesity. *Periodontol. 2000* **2020**, *83*, 125–153. [CrossRef]

78. Isola, G.; Polizzi, A.; Iorio-Siciliano, V.; Alibrandi, A.; Ramaglia, L.; Leonardi, R. Effectiveness of a nutraceutical agent in the non-surgical periodontal therapy: A randomized, controlled clinical trial. *Clin. Oral Investig.* **2020**, 1–11. [CrossRef]

79. Isola, G.; Polizzi, A.; Santonocito, S.; Alibrandi, A.; Ferlito, S. Expression of Salivary and Serum Malondialdehyde and Lipid Profile of Patients with Periodontitis and Coronary Heart Disease. *Int. J. Mol. Sci.* **2019**, *20*, 6061. [CrossRef]

80. Isola, G.; Giudice, A.L.; Polizzi, A.; Alibrandi, A.; Patini, R.; Ferlito, S. Periodontitis and Tooth Loss Have Negative Systemic Impact on Circulating Progenitor Cell Levels: A Clinical Study. *Genes* **2019**, *10*, 1022. [CrossRef]

Article

Oral Microbiome in Four Female Centenarians

Yoshiaki Nomura [1,*], **Erika Kakuta** [2], **Ayako Okada** [1], **Ryoko Otsuka** [1], **Mieko Shimada** [3], **Yasuko Tomizawa** [4], **Chieko Taguchi** [5], **Kazumune Arikawa** [5], **Hideki Daikoku** [6], **Tamotsu Sato** [6] and **Nobuhiro Hanada** [1]

[1] Department of Translational Research, Tsurumi University School of Dental Medicine, Kanagawa 230-8501, Japan; okada-a@tsurumi-u.ac.jp (A.O.); otsuka-ryoko@tsurumi-u.ac.jp (R.O.); hanada-n@tsurumi-u.ac.jp (N.H.)

[2] Department of Oral bacteriology, Tsurumi University School of Dental Medicine, Kanagawa 230-8501, Japan; kakuta-erika@tsurumi-u.ac.jp

[3] Chiba Prefecture University of Health Sciences, Chiba 261-0014, Japan; mieko.shimada@cpuhs.ac.jp

[4] Department of Cardiovascular Surgery, Tokyo Women's Medical University, Tokyo 162-8666, Japan; tomizawa.yasuko@twmu.ac.jp

[5] Department of Preventive and Public Oral Health, Nihon University School of Dentistry at Matsudo, Matsudo 271-8587, Japan; taguchi.chieko@nihon-u.ac.jp (C.T.); arikawa.kazumune@nihon-u.ac.jp (K.A.)

[6] Iwate Dental Association, Morioka 470-2101, Japan; dai-koku@nifty.com (H.D.); tamosato-dent@k-2inc.jp (T.S.)

* Correspondence: nomura-y@tsurumi-u.ac.jp; Tel.: +81-45-580-8462

Received: 29 June 2020; Accepted: 29 July 2020; Published: 31 July 2020

Abstract: The oral microbiome of healthy older adults has valuable information about a healthy microbiome. In this study, we collected and analyzed the oral microbiome of denture plaque and tongue coating samples from four female centenarians. After DNA extraction and purification, pyrosequencing of the V3–V4 hypervariable regions of the 16S rRNA was carried out. The bacterial taxonomy for each lead was assigned based on a search of the EzBioCloud 16S database. We obtained a total of 199,723 valid, quality-controlled reads for denture plaque and 210,750 reads for tongue coating. The reads were assigned 407 operational taxonomic units with a 97% identity cutoff. Twenty-nine species were detected in both denture plaque and tongue coatings from all subjects. *Firmicutes* was the most abundant phylum; the *Streptococcus salivarius* group was the most abundant species in both the denture plaque and tongue coatings; and the *Fusobacterium nucleatum* group was detected in all subjects. In the bacterial profile, species formed clusters composed of bacteria with a wide range of prevalence and abundance, not dependent on phyla; each cluster may have specific species that could be candidates for a core microbiome. *Firmicutes* and *Veillonella* were abundant phyla on both plaque and tongue coatings of centenarians.

Keywords: oral microbiome; core microbiome; centenarian; next-generation sequencing

---

1. Introduction

The human microbiome changes with age [1] and health affects the composition of the microflora. It has been suggested that aging is accompanied by an underlying inflammatory state [2] that interacts with the microbiota of older adults and makes them more susceptible to age-related diseases [3–5]. Changes in the gut microbiome are explained by diseases, including metabolic changes and inflammatory conditions. Studies of intestinal microbiota in the elderly show that the microbiome affects a variety of clinical problems, including physical weakness, *Clostridium difficile* infection, colitis, vulvar vaginal atrophy, colorectal cancer, and atherosclerosis [6]. Other research has focused on the metabolism of nutrients. The microbiota has been shown to correlate with the declining metabolism of essential amino acids by aging [7].

In addition to the intestinal microbiome, the oral microbiome has been intensively studied. Both are affected by development, aging, and the state of oral disease. The formation of the first oral flora is strongly influenced by the mother [8–10]. Breast milk provides a source of bacteria that act as an inoculum for newborns [11–14]. The production and excretion of metabolites by pioneering colonies, such as *Streptococcus* and *Actinomyces*, alter the anaerobic oral environment. Under such conditions, anaerobic bacteria such as *Veillonella* and fusobacteria colonize [8,15]. With development, the microbial community evolves and microbial diversity increases [16,17]. Current knowledge shows that adult-like stability is reached around age two [15]. The established oral microbiota is disturbed by oral diseases. The development of dental caries has been associated with changes in microbial composition over time [18]. Differences in bacterial communities on the phylogenetic level were observed between healthy people and patients with periodontal disease [19,20]. The oral microbiota includes some pathogenic bacteria for systemic diseases. Therefore, the oral microbiome affects health and varies with health or disease [21]. Oral diseases, especially periodontal disease, can be the main risk factor for several systemic diseases [22–27]. Periodontal bacteria and their surface lipopolysaccharides were the agents of systemic diseases [28]. Recent advances of research on systemic diseases and oral health showed that there exist protective host factors for systemic diseases in relation to periodontal diseases [29–31]. The number of studies on the oral microbiota of community-dwelling older persons is limited [32–37].

In this study, denture plaque and tongue surface samples were obtained from people over 100 years old. The samples were analyzed by high-throughput sequencing of 16S rRNA through a metagenomics approach. We describe the commonly prevalent and highly abundant species of four female centenarians and compare them with the healthy oral microbiome proposed in previous studies. In addition, differences in sampling sites and the co-prevalence of species were analyzed.

2. Materials and Methods

2.1. Subject and Setting

Originally, the aim of the survey was to investigate the relationship between oral health and systemic health in 80-year-old adults. This survey is known as the 8020 Data Bank Survey. A 20-year follow-up study was conducted from 1996 to 2017 with subjects 80 years of age (born in 1917) residing in the 8 districts served by one health center in Iwate Prefecture. The sampling method was cluster sampling, and the sampling frame was a complete count survey for all subjects. For the baseline survey, based on residential registration, public health nurses visited 944 homes where 80-year-olds lived, and 666 subjects participated in these checkups.

The follow-up survey was conducted by the resident register with surviving subjects who participated in the baseline survey. After 20 years, 12 subjects survived. Among them, 5 lived in their own home with their families; the others lived in nursing homes. Four subjects who lived in their own home agreed to participate in the survey. We visited their homes at 10:00 in the morning to minimize the effect of food intake. Details of the survey were described in our previous report [27,28,37,38].

2.2. Oral Examination

Two dentists visited the homes of the subjects. An oral examination was carried out with a penlight and a dental mirror. The definition and diagnosis of dental caries were based on the criteria of the World Health Organization [39].

2.3. Sample Collection

As all subjects wore complete dentures, plaque samples were collected from the denture surface based on a method described in previous reports [40,41]. Briefly, the denture was brushed around the buccal surface with a toothbrush for 2 min, followed by immersion of the toothbrush with the attached plaque in sterilized, phosphate-buffered saline (PBS). Tongue coating samples were collected by using a mucosal brush (ERAC 510Lion, Tokyo, Japan). The tongue surface was brushed one way, from back to front, 5 times. Samples were treated in the same way as denture plaque samples. Samples were kept on ice before being transported to the Iwate Dental Association in an ice box with refrigerant, then were stored at −20 °C for further analysis.

2.4. Microbial DNA Extraction

Denture plaque and tongue surface samples suspended in PBS were collected by a centrifuge at 3000 rpm for 10 min. DNA extraction was performed by a Maxwell 16 LEV Blood DNA Kit (Promega KK, Tokyo, Japan)according to the manufacturer's instructions. DNA concentration was measured by a NanoDrop ND-2000 (Thermo Fisher Scientific KK, Tokyo, Japan). Degradation of DNA was visually checked by electrophoresis on 1% agarose gel. Degradation of DNA and contamination of RNA were checked by a Qubit dsDNA BR Assay Kit (Thermo Fisher Scientific KK, Tokyo, Japan).

Samples meeting the following criteria were used for further sequence analysis: conc > 20 ng/μL, volume ≥ 20 μL, A260/280 ≥ 1.8, and A260/230 > 1.5. In this study, all samples met the criteria.

2.5. Microbial Community Analysis

Extracted DNA was analyzed in a laboratory (Chun Lab, Seoul, Korea). Polymerase chain reaction (PCR) amplification was performed using primers specific to the V3–V4 region pyrosequencing tags of the 16S rRNA gene in the extracted bacterial DNA. Taxonomic classification of each read was assigned based on a search of the EzBioCloud 16S database [42,43], which contains the 16S rRNA genes of type strains that have valid published names and representative species-level phylotypes of both cultured and uncultured entries in the GenBank database, with complete hierarchical taxonomic classification from the phylum to species level [44].

2.6. Bioinformatics Analysis

The number of 16S rRNA gene copies (absolute abundance) of operational taxonomic units (OTUs) was calculated by multiplying their respective relative abundance by the total number of 16S rRNA gene copies. Bioinformatics analysis was performed by the microbiome package on the Bioconductor of R software [45].

3. Results

3.1. Characteristics of Subjects Who Participated in this Study

All four subjects who participated in this study were women, 100 years old, and wore complete dentures. One subject had five residual roots (sample ID 1), with complete dentures designed over the stump root. Blood tests and medical examinations were conducted to confirm the health of the subjects. Blood tests showed values in the normal range. Two subjects had hypertension. According to this continuous study, one subject had high blood pressure at age 85 (sample ID 1) and one had normal blood pressure at age 90 (sample ID 2). Other than that, there were no specific symptoms on examination.

3.2. Sequence Data Details

Out of eight samples from the four subjects, 199,723 reads for denture plaque (min, max: 42,792, 54,898) and 210,750 for tongue coatings (45,193, 57,753) passed quality control. From these reads, sequences clustered into 15 phyla, 31 classes, 49 orders, 71 families, 142 genera, and 406 species. A heatmap of all 407 species detected in this study is shown in Figure S1. All sequence data are provided in the Supplementary Materials.

The ACE, Chao1, jackknife, Shannon, and Simpson alpha diversity indices were calculated to analyze the diversity and richness of all samples. By comparing the samples of denture plaque and tongues, the ACE, Chao1, jackknife, and Shannon indices were not significantly different ($P > 0.05$), proving that bacterial diversity and richness were similar in the samples collected from denture plaque and tongues. The values of these indices are shown in Table S1. Statistics of taxonomic assignment are shown in Table S2. A rarefaction curve is shown in Figure S2.

3.3. Oral Microbiome Profile of Centenarians

Figure 1A shows the relative abundance of detected bacteria at the phylum level. On both denture plaque and the tongue, Firmicutes were abundant. Figure 1B shows the phylum level proportion of bacteria of the four subjects. For three subjects, Firmicutes were abundant in both saliva and the tongue. Proteobacteria were relatively abundant in the saliva and tongue of one subject. Dynamic pie charts are presented in the Supplementary Materials.

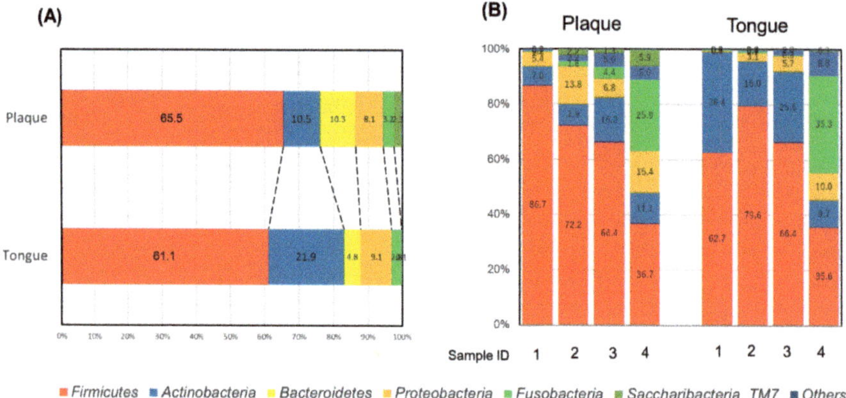

Figure 1. Proportion of phyla in denture plaque and the tongue: (**A**) total, and (**B**) individual subjects.

3.4. Core Microbiome Analysis

Figure 2 shows the prevalence of taxa in terms of abundance. Abundant species belonged to *Firmicutes*, *Proteobacteria*, and *Bacteroides*. In this study, 29 species were detected from all subjects in denture plaque and the tongue. Table 1 shows the species detected from all subjects, and Table S3 shows the 15 species in plaque and 7 species in the tongue. The most abundant bacteria were Firmicutes at the phylum level and the *Streptococcus salivarius* group at the species level in both plaque and the tongue. The *Fusobacterium nucleatum* group was detected from all subjects. A heatmap of the candidates for the core microbiome is shown in Figure S3.

Table 1. Bacteria detected in denture plaque and tongue from all subjects.

Phylum	Genus	Species	Abundance (%)	
			Plaque	**Tongue**
Firmicutes	Streptococcus	*Streptococcus salivarius* group	9.72	25.06
		Streptococcus sinensis group	4.12	13.22
		Streptococcus pneumoniae group	4.46	1.92
		Streptococcus parasanguinis group	2.30	3.83
		Streptococcus gordonii group	4.80	0.62
		Streptococcus peroris group	0.82	1.03
		Streptococcus sanguinis	0.87	0.08
		Streptococcus_uc	0.40	0.32
	Veillonella	*Veillonella dispar*	5.63	4.89
		Veillonella parvula group	7.40	0.51
		Veillonella atypica	1.90	1.95
		Veillonella_uc	0.05	0.09
	Gemella	*Gemella haemolysans* group	0.23	0.54
	Granulicatella	*Granulicatella adiacens* group	0.43	1.67
	Lachnoanaerobaculum	*Lachnoanaerobaculum orale* group	0.13	0.08
	Lactobacillus	*Lactobacillus salivarius*	3.95	0.33
	Megasphaera	*Megasphaera micronuciformis*	0.29	0.07
	Moryella	*Stomatobaculum longum*	0.17	0.22
Actinobacteria	Actinomyces	*KE952139_s*	0.71	2.39
		CAGY_s	0.11	0.30
		JRMV_s	0.02	0.37
	Atopobium	*Atopobium parvulum*	0.76	0.27
	Rothia	*Rothia mucilaginosa*	0.65	17.17
		Rothia dentocariosa	4.34	0.52
		Rothia_uc	0.02	0.10
Bacteroidetes	Prevotella	*Prevotella histicola*	3.95	0.54
		Prevotella jejuni	0.32	1.06
		Prevotella salivae	0.35	0.08
Fusobacteria	Fusobacterium	*Fusobacterium nucleatum* group	0.37	0.04

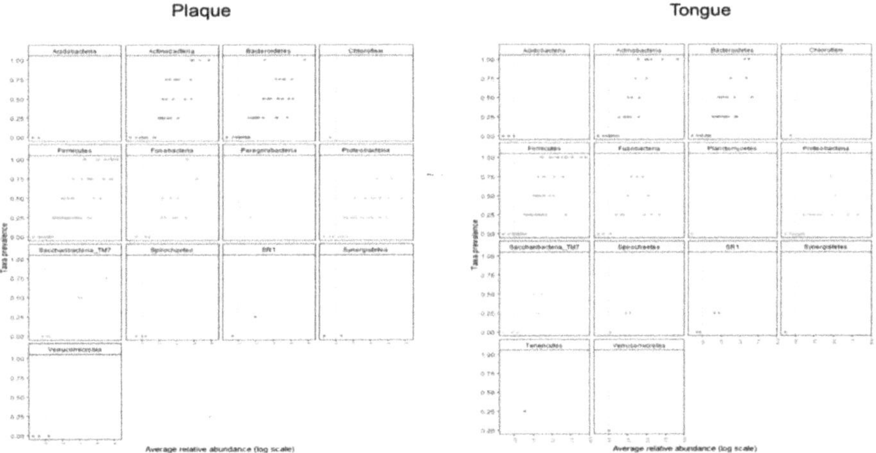

Figure 2. Prevalence of taxa in terms of abundance for 14 different phyla.

3.5. Analysis of Co-prevalent Species

Then, the levels of species co-prevalence were analyzed by t-distributed stochastic neighbor embedding (tSNE). As shown in Figure 3, species were constructed in clusters that were not dependent on the phylum. The list of classified species and detection in 3 of 4 subjects (75%) is shown in Table S2.

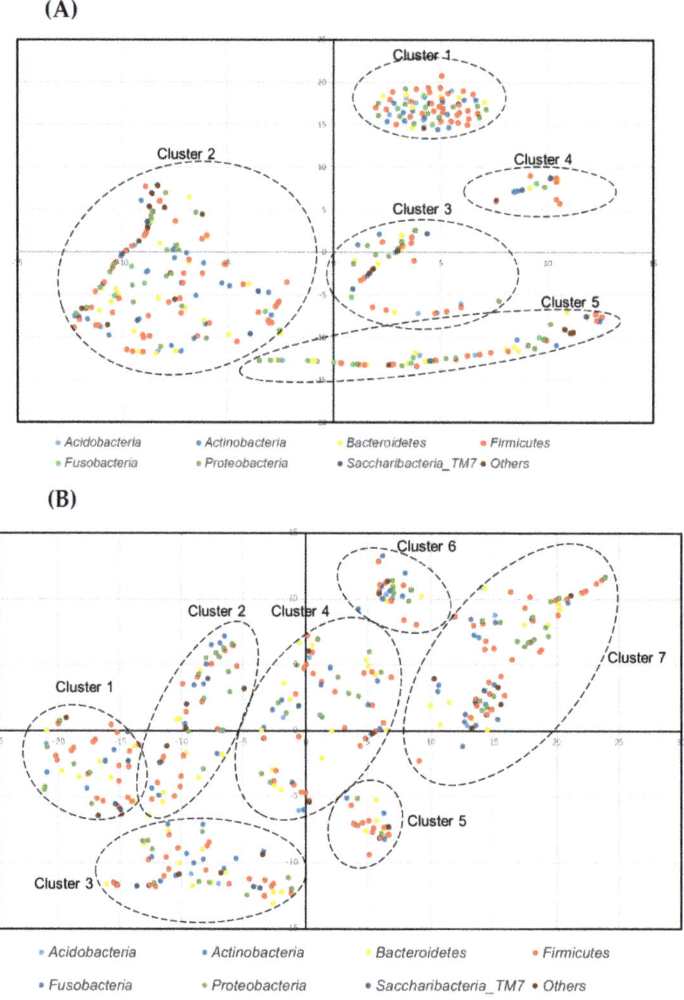

Figure 3. t-distributed stochastic neighbor embedding (tSNE) plot of species: (**A**) denture plaque, and (**B**) the tongue.

4. Discussion

This study describes microbial profiles of the oral cavities of four female centenarians by high throughput sequencing. Highly prevalent and abundant species are described. In addition, differences in sampling sites and co-prevalence of species are presented.

A total of 406 species were detected. Studies indicate that the diversity of bacteria is reduced with the frailty of older adults [3,46]. All participants in this study were female centenarians. Bacterial diversity was not reduced when compared to another study [16].

Firmicutes was the most abundant phylum for both the denture plaque and tongue samples. Several studies have shown that *Firmicutes* is the most abundant phylum in the saliva of infants [16] and young children aged 6–8 years [47], swabs of the oral cavities of youths and adults [48], and dental plaque from older adults [34]. *Firmicutes, Streptococcus,* and *Veillonella* were detected in all subjects in both the denture plaque and the tongue. The result that *Firmicutes* was the most abundant, followed by *Veillonella,* is consistent with the proportions of oral microorganisms in the Human Oral Microbiome Database [49]. The abundant genera after *Veillonella* vary between studies. *Streptococcus, Veillonella,* and *Neisseria* were the predominant bacterial genera present in infants. Abundant genera that coat the tooth surface first are *Streptococcus* and *Veillonella* [50]. *Streptococcus, Veillonella,* and *Neisseria* are the predominant bacterial genera present in infants [16]. *Streptococcus* and *Veillonella* are known as early colonizers. *Neisseria* settles down at the age of 1–2 years [18].

In this study, *Neisseria* was not detected in all subjects. Differences at the species level in previous reports may be due to the sample difference. All subjects surveyed in this study wore complete dentures, so the plaque sample was denture plaque, not dental plaque attached to the tooth surface. Additionally, the small sample size may have affected the difference.

In this study, *Streptococcus salivarius* was the most abundant species in both the denture plaque and the tongue. This result is consistent with another study that investigated the tongues of Japanese older adults [16]. The results of the study show that more than 5% of abundant species were *Streptococcus salivarius, Prevotella melaninogenica, Rothia mucilaginosa, Veillonella atypica,* and *Neisseria flavescens.* Except for *Neisseria flavescens,* the other four species were common to our results. Another study showed that *Streptococcus oralis* was by far the most prevalent species [51]. As described above, Firmicutes and *Veillonella* are abundant phyla. Each study has different bacterial profiles at the genus or species level.

The bacterial profile of the denture plaque was similar to that of oral mucosa. Actinobacteria was abundant in the denture plaque and Bacilli were abundant in the dental plaque [52]. In this study, the proportion of these bacteria varies between samples: 5.4%, 15.5%, 10.8%, and 6.5% for Actinobacteria and 82.45%, 44.4%, 21.5%, and 48.1% for Bacilli. The small sample size may have affected the results.

A previous study had shown that oral microbiome profiles were related to the risk of death by pneumonia for older persons residing in nursing homes [53]. *Neisseria flavescens,* the *Fusobacterium periodonticum* group, and the *Haemophilus parainfluenzae* group were abundant in the risk group. In this study, one subject (Sample ID 4) had higher levels of these bacteria than the mean values of the risk group.

As shown in Figure 3 the bacterial community formed clusters with a low prevalence or abundance of bacteria. The clustering was not phylum dependent.

The clusters were composed of bacteria with a wide range of prevalence and abundance. This indicates that some bacteria were the core of the cluster. There may be an interaction or symbiotic mechanism between core bacteria and low-prevalence or low-abundance bacteria. Therefore, there may be healthy microbiome candidates within the core bacteria of each cluster.

A limitation of this study was the small sample size of four people. The design was a cross-sectional study. The data presented in this study were from Japanese older persons. Even though dietary habits have little effect on the oral microbiome [54], regional or cultural differences could have an effect on the oral microbiome [44]. The health status of the subjects who participated in this study was limited. Subjects with specific diseases were not included. For comparison, healthy and pathogenic conditions should be considered. Longevity is not simply explained by microflora. Further study is needed to evaluate the host factors such as nutraceutical agent [29], serum malondialdehyde [30], and progenitor cell levels [31]. In addition, the subjects investigated in this study were all female and edentulous.

5. Conclusions

The oral microbiome was preserved in Japanese centenarians at the phylum level. At the species level, bacterial profiles were not consistent with other studies. *Firmicutes* and *Veillonella* were abundant phyla in both the plaque and the tongue. In the bacterial profile, bacteria formed clusters. The oral microbiome of the centenarian investigated in this study was similar to that of other age groups from previous studies at phylum level. Further study is needed to define a common bacterial profile at the species level.

Supplementary Materials: The following are available online at http://www.mdpi.com/2076-3417/10/15/5312/s1. Figure S1: Heatmap of all species detected. Figure S2: Rarefaction curve. Figure S3: Core heatmap. Table S1: Alpha diversity indices of different groups. Table S2: Statistics of taxonomic assignment. Table S3: Species detected in denture plaque and tongue. Sequence data: Sequence data of the eight samples. Dynamic pie chart: Dynamic pie chart of the bacterial profiles of samples. OTU data: OTU data used in this study.

Author Contributions: Y.N. planned the study design and management and analysis of the data and wrote the original draft. E.K., A.O., R.O., M.S., Y.T., C.T., and K.A. collected and managed the data. Y.N., H.D., T.S., and N.H. contributed to funding acquisition, planning the study design, and reviewing and editing the manuscript. All authors have read and agreed to the published version of the manuscript.

Funding: This study was supported by JSPS KAKENHI (grant numbers 17K12030, 20K10303), SECOM Science and Technology Foundation, and an 8020 Research Grant for fiscal year 2017 from the 8020 Promotion Foundation (grant number 17–2–05). None of the funders played a role in the design of the study, data collection, analysis, interpretation of the results, or writing of the manuscript.

Conflicts of Interest: The authors state that they have no financial or nonfinancial conflict of interest regarding this research.

References

1. Yatsunenko, Y.; Rey, F.E.T.; Manary, M.J.; Trehan, I.; Dominguez-Bello, M.G.; Contreras, M.; Magris, M.; Hidalgo, G.; Baldassano, R.N.; Anokhin, A.P.; et al. Human gut microbiome viewed across age and geography. *Nature* **2012**, *486*, 222–227. [CrossRef]

2. Chung, H.Y.; Lee, E.K.; Choi, Y.J.; Kim, J.M.; Kim, D.H.; Zou, Y.; Kim, C.H.; Lee, J.; Kim, H.S.; Kim, N.D.; et al. Molecular inflammation as an underlying mechanism of the aging process and age-related diseases. *J. Dent. Res.* **2011**, *90*, 830–840. [CrossRef] [PubMed]

3. Claesson, M.J.; Cusack, S.; O'Sullivan, O.; Greene-Diniz, R.; de Weerd, H.; Flannery, E.; Marchesi, J.R.; Falush, D.; Dinan, T.; Fitzgerald, G.; et al. Composition, variability, and temporal stability of the intestinal microbiota of the elderly. *Proc. Natl. Acad. Sci. USA* **2011**, *108*, 4586–4591. [CrossRef]

4. O'Toole, P.W.; Jeffery, I.B. Gut microbiota and aging. *Science* **2015**, *350*, 1214–1215. [CrossRef]

5. Jackson, M.A.; Jeffery, I.B.; Beaumont, M.; Bell, J.T.; Clark, A.G.; Ley, R.E.; O'Toole, P.W.; Spector, T.D.; Steves, C.J. Signatures of early frailty in the gut microbiota. *Genome Med.* **2016**, *8*, 8. [CrossRef]

6. Zapata, H.J.; Quagliarello, V.J. The microbiota and microbiome in aging: Potential implications in health and age-related diseases. *J. Am. Geriatr. Soc.* **2015**, *63*, 776–781. [CrossRef]

7. Rampelli, S.; Candela, M.; Turroni, S.; Biagi, E.; Collino, S.; Franceschi, C.; O'Toole, P.W.; Brigidi, P. Functional metagenomic profiling of intestinal microbiome in extreme ageing. *Aging* **2013**, *5*, 902–912. [CrossRef]

8. Sampaio-Maia, B.; Monteiro-Silva, F. Acquisition and maturation of oral microbiome throughout childhood: An update. *Dent. Res. J.* **2014**, *11*, 291–301.

9. Flores, G.E.; Caporaso, J.G.; Henley, J.B.; Rideout, J.R.; Domogala, D.; Chase, J.; Leff, J.W.; Vázquez-Baeza, Y.; Gonzalez, A.; Knight, R.; et al. Temporal variability is a personalized feature of the human microbiome. *Genome Biol.* **2014**, *15*, 531. [CrossRef] [PubMed]

10. Teng, F.; Yang, F.; Huang, S.; Bo, C.; Xu, Z.Z.; Amir, A.; Knight, R.; Ling, J.; Xu, J. Prediction of early childhood caries via spatial-temporal variations of oral microbiota. *Cell Host Microbe* **2015**, *18*, 296–306. [CrossRef] [PubMed]

11. Boix-Amorós, A.; Collado, M.C.; Mira, A. Relationship between milk microbiota, bacterial load, macronutrients, and human cells during lactation. *Front Microbiol.* **2016**, *7*, 492. [CrossRef] [PubMed]

12. Fernández, L.; Langa, S.; Martín, V.; Maldonado, A.; Jiménez, E.; Martín, R.; Rodríguez, J.M. The human milk microbiota: Origin and potential roles in health and disease. *Pharmacol. Res.* **2013**, *69*, 1–10. [CrossRef] [PubMed]

13. Fitzstevens, J.L.; Smith, K.C.; Hagadorn, J.I.; Caimano, M.J.; Matson, A.P.; Brownell, E.A. Systematic review of the human milk microbiota. *Nutr. Clin. Pract.* **2016**, *32*, 354–364. [CrossRef] [PubMed]

14. Rodriguez, J.M. The origin of human milk bacteria: Is there a bacterial entero-mammary pathway during late pregnancy and lactation? *Adv. Nutr. Int. Rev. J.* **2014**, *5*, 779–784. [CrossRef]

15. Gomez, A.; Nelson, K.E. The oral microbiome of children: Development, disease, and implications beyond oral health. *Microb. Ecol.* **2017**, *73*, 492–503. [CrossRef]

16. Cephas, K.D.; Kim, J.; Mathai, R.A.; Barry, K.A.; Dowd, S.E.; Meline, B.S.; Swanson, K.S. Comparative analysis of salivary bacterial microbiome diversity in edentulous infants and their mothers or primary care givers using pyrosequencing. *PLoS ONE* **2011**, *6*, e23503. [CrossRef]

17. Lif Holgerson, P.; Öhman, C.; Rönnlund, A.; Johansson, I. Maturation of oral microbiota in children with or without dental caries. *PLoS ONE* **2015**, *10*, e0128534. [CrossRef]

18. Dzidic, M.; Collado, M.C.; Abrahamsson, T.; Artacho, A.; Stensson, M.; Jenmalm, M.C.; Mira, A. Oral microbiome development during childhood: An ecological succession influenced by postnatal factors and associated with tooth decay. *ISME J.* **2018**, *12*, 2292–2306. [CrossRef]

19. Chen, W.P.; Chang, S.H.; Tang, C.Y.; Liou, M.L.; Tsai, S.J.; Lin, Y.L. Composition Analysis and Feature Selection of the Oral Microbiota Associated with Periodontal Disease. *Biomed. Res. Int.* **2018**, *2018*, 3130607. [CrossRef]

20. Minty, M.; Canceil, T.; Serino, M.; Burcelin, R.; Tercé, F.; Blasco-Baque, V. Oral microbiota-induced periodontitis: A new risk factor of metabolic diseases. *Rev. Endocr. Metab. Disord.* **2019**, *20*, 449–459. [CrossRef]

21. Graves, D.T.; Corrêa, J.D.; Silva, T.A. The Oral Microbiota Is Modified by Systemic Diseases. *J. Dent Res.* **2019**, *98*, 148–156. [CrossRef] [PubMed]

22. Jensen, E.; Allen, G.; Bednarz, J.; Couper, J.; Peña, A. Periodontal risk markers in children and adolescents with type 1 diabetes: A systematic review and meta-analysis. *Diabetes Metab. Res. Rev.* **2020**, e3368. [CrossRef] [PubMed]

23. Wang, J.; Yang, X.; Zou, X.; Zhang, Y.; Wang, J.; Wang, Y. Relationship between periodontal disease and lung cancer: A systematic review and meta-analysis. *J. Periodontal Res.* **2020**. [CrossRef]

24. Nadim, R.; Tang, J.; Dilmohamed, A.; Yuan, S.; Wu, C.; Bakre, A.T.; Partridge, M.; Ni, J.; Copeland, J.R.; Anstey, K.J.; et al. Influence of periodontal disease on risk of dementia: A systematic literature review and a meta-analysis. *Eur. J. Epidemiol.* **2020**. [CrossRef] [PubMed]

25. Priyamvara, A.; Dey, A.K.; Bandyopadhyay, D.; Katikineni, V.; Zaghlol, R.; Basyal, B.; Barssoum, K.; Amarin, R.; Bhatt, D.L.; Lavie, C.J. Periodontal Inflammation and the Risk of Cardiovascular Disease. *Curr. Atheroscler. Rep.* **2020**, *22*, 28. [CrossRef] [PubMed]

26. Orlandi, M.; Graziani, F.; D'Aiuto, F. Periodontal therapy and cardiovascular risk. *Periodontol 2000* **2020**. [CrossRef]

27. Jepsen, S.; Suvan, J.; Deschner, J. The association of periodontal diseases with metabolic syndrome and obesity. *Periodontol 2000* **2020**. [CrossRef]

28. Mulhall, H.; Huck, O.; Amar, S. Porphyromonas gingivalis, a Long-Range Pathogen: Systemic Impact and Therapeutic Implications. *Microorganisms* **2020**, *8*, 869. [CrossRef]

29. Isola, G.; Polizzi, A.; Iorio-Siciliano, V.; Alibrandi, A.; Ramaglia, L.; Leonardi, R. Effectiveness of a nutraceutical agent in the non-surgical periodontal therapy: A randomized, controlled clinical trial. *Clin. Oral Investig.* **2020**. [CrossRef]

30. Isola, G.; Polizzi, A.; Santonocito, S.; Alibrandi, A.; Ferlito, S. Expression of Salivary and Serum Malondialdehyde and Lipid Profile of Patients with Periodontitis and Coronary Heart Disease. *Int. J. Mol. Sci.* **2019**, *20*, 6061. [CrossRef]

31. Isola, G.; Giudice, A.L.; Polizzi, A.; Alibrandi, A.; Patini, R.; Ferlito, S. Periodontitis and Tooth Loss Have Negative Systemic Impact on Circulating Progenitor Cell Levels: A Clinical Study. *Genes* **2019**, *10*, 1022. [CrossRef]

32. Asakawa, M.; Takeshita, T.; Furuta, M.; Kageyama, S.; Takeuchi, K.; Hata, J.; Ninomiya, T.; Yamashita, Y. Tongue Microbiota and Oral Health Status in Community-Dwelling Elderly Adults. *mSphere* **2018**, *3*. [CrossRef] [PubMed]

33. Feres, M.; Teles, F.; Teles, R.; Figueiredo, L.C.; Faveri, M. The subgingival periodontal microbiota of the aging mouth. *Periodontol 2000* **2016**, *72*, 30–53. [CrossRef] [PubMed]
34. Jiang, Q.; Liu, J.; Chen, L.; Gan, N.; Yang, D. The Oral Microbiome in the Elderly with Dental Caries and Health. *Front. Cell Infect. Microbiol.* **2019**, *8*, 442. [CrossRef] [PubMed]
35. Iwauchi, M.; Horigome, A.; Ishikawa, K.; Mikuni, A.; Nakano, M.; Xiao, J.Z.; Odamaki, T.; Hironaka, S. Relationship between oral and gut microbiota in elderly people. *Immun. Inflamm. Dis.* **2019**, *7*, 229–236. [CrossRef] [PubMed]
36. Ogawa, T.; Hirose, Y.; Honda-Ogawa, M.; Sugimoto, M.; Sasaki, S.; Kibi, M.; Kawabata, S.; Ikebe, K.; Maeda, Y. Composition of salivary microbiota in elderly subjects. *Sci. Rep.* **2018**, *8*, 414. [CrossRef] [PubMed]
37. Zakaria, M.N.; Furuta, M.; Takeshita, T.; Shibata, Y.; Sundari, R.; Eshima, N.; Ninomiya, T.; Yamashita, Y. Oral microbiome in community-dwelling elderly and its relation to oral and general health conditions. *Oral Dis.* **2017**, *23*, 973–982. [CrossRef]
38. Nomura, Y.; Kakuta, E.; Okada, A.; Otsuka, R.; Shimada, M.; Tomizawa, Y.; Taguchi, C.; Arikawa, K.; Daikoku, H.; Sato, T.; et al. Effects of self-assessed chewing ability, tooth loss and serum albumin on mortality in 80-year-old individuals: A 20-year follow-up study. *BMC Oral Health.* **2020**, *20*, 122. [CrossRef]
39. World Health Organization. *Oral Health Surveys: Basic Methods*, 5th ed.; World Health Organization: Geneva, Switzerland, 2013.
40. Okada, M.; Hayashi, F.; Nagasaka, N. PCR detection of 5 Putative periodontal pathogens in dental plaque samples from children 2 to 12 years of age. *J. Clin. Periodontol.* **2001**, *28*, 576–582. [CrossRef]
41. Okada, A.; Sogabe, K.; Takeuchi, H.; Okamoto, M.; Nomura, Y.; Hanada, N. Characterization of specimens obtained by different sampling methods for evaluation of periodontal bacteria. *J. Oral Sci.* **2017**, *59*, 491–498. [CrossRef]
42. Kim, O.S.; Cho, Y.J.; Lee, K.; Yoon, S.H.; Kim, M.; Na, H.; Park, S.C.; Jeon, Y.S.; Lee, J.H.; Yi, H.; et al. Introducing EzTaxon-e: A prokaryotic 16S rRNA gene sequence database with phylotypes that represent uncultured species. *Int. J. Syst. Evol. Microbiol.* **2012**, *62*, 716–721. [CrossRef] [PubMed]
43. Yoon, S.H.; Ha, S.M.; Kwon, S.; Lim, J.; Kim, Y.; Seo, H.; Chun, J. Introducing EzBioCloud: A taxonomically united database of 16S rRNA and whole genome assemblies. *Int. J. Syst. Evol. Microbiol.* **2017**, *67*, 1613–1617. [CrossRef]
44. Nomura, Y.; Otsuka, R.; Hasegawa, R.; Hanada, N. Oral Microbiome of Children Living in an Isolated Area in Myanmar. *Int. J. Environ. Res. Public Health* **2020**, *17*, 4033. [CrossRef] [PubMed]
45. Lahti, L.; Shetty, S. Introduction to the Microbiome R Package, 2020. Available online: https://microbiome.github.io/tutorials/ (accessed on 31 July 2020).
46. Saraswati, S.; Sitaraman, R. Aging and the human gut microbiota-from correlation to causality. *Front. Microbiol.* **2014**, *5*, 764. [CrossRef] [PubMed]
47. Xu, Y.; Jia, Y.H.; Chen, L.; Huang, W.M.; Yang, D.Q. Metagenomic analysis of oral microbiome in young children aged 6–8 years living in a rural isolated Chinese province. *Oral Dis.* **2018**, *24*, 1115–1125. [CrossRef]
48. Burcham, Z.M.; Garneau, N.L.; Comstock, S.S.; Tucker, R.M.; Knight, R.; Metcalf, J.L. Patterns of Oral Microbiota Diversity in Adults and Children: A Crowdsourced Population Study. *Sci. Rep.* **2020**, *10*, 2133. [CrossRef]
49. Forsyth Institutes. eHOMED Expanded Human Oral Microbiome Database. Available online: http://www.homd.org/ (accessed on 31 July 2020).
50. Kolenbrander, P.E.; Palmer, R.J.; Periasamy, S.; Jakubovics, N.S. Oral multispecies biofilm development and the key role of cell-cell distance. *Nat. Rev. Microbiol.* **2010**, *8*, 471–480. [CrossRef]
51. Aas, J.A.; Paster, B.J.; Stokes, L.N.; Olsen, I.; Dewhirst, F.E. Defining the normal bacterial flora of the oral cavity. *J. Clin. Microbiol.* **2005**, *43*, 5721–5732. [CrossRef]
52. O'Donnell, L.E.; Robertson, D.; Nile, C.J.; Cross, L.J.; Riggio, M.; Sherriff, A.; Bradshaw, D.; Lambert, M.; Malcolm, J.; Buijs, M.J.; et al. The Oral Microbiome of Denture wearers is influenced by levels of natural dentition. *PLoS ONE* **2015**, *10*, e0137717. [CrossRef]

53. Kageyama, S.; Takeshita, T.; Furuta, M.; Tomioka, M.; Asakawa, M.; Suma, S.; Takeuchi, K.; Shibata, Y.; Iwasa, Y.; Yamashita, Y. Relationships of Variations in the Tongue Microbiota and Pneumonia Mortality in Nursing Home Residents. *J. Gerontol. A Biol. Sci. Med. Sci.* **2018**, *73*, 1097–1102. [CrossRef]
54. De Filippis, F.; Vannini, L.; LaStoria, A.; Laghi, L.; Piombino, P.; Stellato, G.; Serrazanetti, D.I.; Gozzi, G.; Turroni, S.; Ferrocino, I.; et al. The same microbiota and a potentially discriminant metabolome in the saliva of omnivore, ovo-lacto-vegetarian and Vegan individuals. *PLoS ONE* **2014**, *9*, e112373. [CrossRef] [PubMed]

applied sciences

Article

Periodontal Condition and Subgingival Microbiota Characterization in Subjects with Down Syndrome

Maigualida Cuenca [1], María José Marín [1], Lourdes Nóvoa [2], Ana O'Connor [1], María Carmen Sánchez [1], Juan Blanco [2], Jacobo Limeres [2], Mariano Sanz [1], Pedro Diz [2,†] and David Herrera [1,*,†]

[1]   ETEP (Etiology and Therapy of Periodontal and Peri-implant Diseases) Research Group, University Complutense of Madrid (UCM), 28040 Madrid, Spain; maiguacu@ucm.es (M.C.); mjmarin@ucm.es (M.J.M.); aoconnor@ucm.es (A.O.); mariasan@ucm.es (M.d.C.S.); marsan@ucm.es (M.S.)

[2]   Medical-Surgical Dentistry Research Group (OMEQUI), Health Research Institute of Santiago de Compostela (IDIS), University of Santiago de Compostela (USC), 15782 Santiago de Compostela, Spain; lourdes.novoa@rai.usc.es (L.N.); juan.blanco@usc.es (J.B.); jacobo.limeres@usc.es (J.L.); pedro.diz@usc.es (P.D.)

*   Correspondence: davidher@ucm.es; Tel.: +34-913-941-907

†   Contributed equally.

Featured Application: **Promotion of preventive actions in Down syndrome individuals is recommended, including surveillance of thyroid hormone function, improvement of oral hygiene measures and frequent evaluation of periodontal health.**

Abstract: The aim was to study the subgingival microbiota in subjects with Down syndrome (DS) with different periodontal health status, using cultural and molecular microbiological methods. In this cross-sectional study, DS subjects were selected among those attending educational or occupational therapy centers in Galicia (Spain). Medical histories, intraoral and periodontal examinations and microbiological sampling were performed. Samples were processed by means of culture and quantitative polymerase chain reaction (qPCR). Microbiological data were compared, by one-way ANOVA or Kruskal-Wallis and chi-square or Fisher tests, according to their periodontal status. 124 subjects were included, 62 with a healthy periodontium, 34 with gingivitis and 28 with periodontitis. Patients with periodontitis were older ($p < 0.01$) and showed lower prevalence of hypothyroidism and levothyroxine intake ($p = 0.01$), presented significantly deeper pockets and more attachment loss ($p \leq 0.01$). Both gingivitis and periodontitis subjects showed higher levels of bleeding and dental plaque. PCR counts of *T. forsythia and* culture counts of *E. corrodens* and total anaerobic counts were significantly higher in periodontitis patients. Relevant differences were observed in the subgingival microbiota of DS patients with periodontitis, showing higher levels of anaerobic bacteria, *T. forsythia* and *E. corrodens*, when compared with periodontally healthy and gingivitis subjects. Moreover, periodontitis subjects were older, had lower frequency of hypothyroidism and higher levels of dental plaque.

Keywords: periodontitis; Down syndrome; polymerase chain reaction; microbiota; gingivitis

Citation: Cuenca, M.; Marín, M.J.; Nóvoa, L.; O'Connor, A.; Sánchez, M.C.; Blanco, J.; Limeres, J.; Sanz, M.; Diz, P.; Herrera, D. Periodontal Condition and Subgingival Microbiota Characterization in Subjects with Down Syndrome. *Appl. Sci.* **2021**, *11*, 778. https://doi.org/10.3390/app11020778

Received: 6 December 2020
Accepted: 11 January 2021
Published: 15 January 2021

1. Introduction

The introduction Down syndrome (DS) is a congenital genetic disease, caused by the presence of an extra chromosome in par 21 [1], being the most common human aneuploidy among living births, with an estimated prevalence of one case per 800 births [2]. Its phenotypic expression, however, may be associated with significant genetic complexity [3]. Subjects with DS present a characteristic phenotype and intellectual disability [4], together with several systemic diseases [5], such as congenital heart disease and thyroid dysfunction, as well as a number of oral diseases [6,7]. Furthermore, there is compromise in their innate immune response [8] and DS subjects suffer more frequently from oral infections, when compared with the general population [9,10].

Early onset forms of periodontitis and late onset of caries lesions represent the most frequent oral conditions associated with DS subjects [9], which may become further aggravated by their motor disability and manual dexterity compromise, that limits their performance in oral hygiene practices [11]. Several observational studies have reported high prevalence of gingivitis and periodontitis (ranging from 58–96%) at young ages (less than 35) in DS subjects [12]. In Brazil, in a study sample of 93 DS patients aged 6–20 years, only 9% showed a healthy periodontium and 33% had periodontitis [13]; in another study including 64 DS subjects (mean age 23.8 years), 28.1% presented gingivitis and 71.9% periodontitis [14]. In The Netherlands, in a study group conformed by 182 DS subjects, 36.6% were diagnosed of periodontitis [15]. In addition, SD periodontitis patients have shown more severity, when compared with controls [16], and a higher impact in their quality of life [13]. In fact, in the latest classification of periodontal and peri-implant diseases and conditions [17], periodontitis in DS patients was included within the category "periodontitis as a manifestation of a systemic disease," in which the systemic condition has a major impact in the onset and course of periodontitis, as clearly occurs in DS subjects [18].

Furthermore, some authors have suggested a specific microbial profile associated to DS subjects with periodontitis [19,20], raising the hypothesis that a distinctive pathogenic microbiota combined with an impaired host-response [21], may result in an earlier dysbiosis and onset of periodontitis [22]. However, the evidence supporting this hypothesis is very scarce and the reported results are highly heterogeneous [19,20,23]. In light of this limited information, we have designed this observational study aimed to evaluate the subgingival microbiota of DS subjects with different periodontal health status (periodontal health, gingivitis or periodontitis), by means of cultural and molecular microbiological methods.

2. Materials and Methods

2.1. Study Population

A cross-sectional observational study was designed in a Spanish Caucasian population diagnosed of DS. This investigation followed the Strengthening the Reporting of Observational studies in Epidemiology (STROBE) criteria for reporting [24] and was approved by the Research Ethics Committee (EC) of Santiago-Lugo, Spain (Registration Code: 2018/510). All subjects and, when appropriate, their legal guardians, confirmed in writing their agreement to participate and signed the EC-approved informed consent, once they were informed on the objectives and processes associated with this investigation.

Screening among DS subjects regularly attending seven educational or occupational therapy centers in the region of Galicia, in the North-West of Spain, was carried out between September 2017 and November 2018.

At this screening visit, information on their age, gender, presence of co-morbidities and use of current medications was collected. Study subjects were consecutively included if DS was genetically confirmed and did not have any of the following exclusion criteria: under 18 years of age, comorbidities that could influence the periodontal condition (e.g., diabetes), presence of harmful habits (e.g., smoking), having received antimicrobial therapy in the previous month (e.g., systemic antimicrobials and/or oral antiseptics), insufficient degree of collaboration for performing clinical assessment and microbiological sampling.

Upon inclusion in the study, subjects had an intraoral examination where the following periodontal variables were recorded in four sites at the six reference teeth [25] or when absent, at the adjacent tooth: plaque index (PlI) [26], probing depth (PD) (using a PCP UNC15 periodontal probe, applying a force of 20–25 g), bleeding on probing (BOP) [27], gingival recession (REC) (distance from the gingival margin to the cemento-enamel junction) and clinical attachment level (CAL). All measurements were recorded by a trained and calibrated clinical investigator (N.L.). Calibration was achieved during several sessions of repeated measurements, until the clinical investigator demonstrated an intra-class correlation coefficient (ICC) higher than 0.75 in all the periodontal variables evaluated. The intra-class correlation coefficient values for intra-examiner calibration were 0.87 (95% confidence interval = 0.75–0.98). Clinical assessments were not conducted in dental clinics/chairs so, together with the specific condition of the subjects, neither full-mouth clinical evaluations or radiographical assessments were feasible.

Following a modification of the case definitions proposed by the Classification of Periodontal and Peri-Implant Diseases and Conditions [17], the study group was categorized as:

- Healthy periodontal condition: PD < 4 mm and BOP detected in no more than two sites (less than 10% of the total) [28],
- Gingivitis: PD < 4 mm and BOP detected in more than two sites (more than 10% of the total) [28],
- Periodontitis: PD $\geq$ 4 mm in, at least, one site [29].

2.2. Subgingival Biofilm Sample Collection

Subgingival samples were collected from four sites, selected by clinical criteria, namely the deepest PD with BOP per quadrant [30]. Sampling was performed after the isolation of the area by cotton rolls and after removing supragingival biofilm and/or calculus. Then, two medium-sized sterile paper tips (Maillefer, Ballaigues, Switzerland) were consecutively inserted in each site for 10 s, as subgingival as possible [30]. All eight paper points were pooled in a vial containing 1.5 mm of reduced transport fluid (RTF) [31] and sent to the Laboratory of Microbiology (Faculty of Dentistry, University Complutense of Madrid, Spain), to be processed within 24 h.

2.3. Microbiological Processing by Means of Quantitative Polymerase Chain Reaction

2.3.1. Extraction of Total Genomic DNA

Total DNA was extracted from subgingival samples using a commercial kit (MolYsis Complete 5, Molzym Gmbh & Co. KG. Bremen, Germany) following manufacturer's instructions (the protocol for bacterial DNA extraction was followed from step 6, avoiding preliminary steps). The extracted DNA was eluded in 100 µL of sterile water (Roche Diagnostic GmbH, Mannheim, Germany) and frozen at -20 °C for further analysis.

2.3.2. Polymerase Chain Reaction

Multiplex quantitative PCR (qPCR) technology was used for detecting and quantifying the bacterial DNA [32]. The sequence of the primers and probes used for *Aggregatibacter actinomycetemcomitans (A. actinomycetemcomitans)*, *Porphyromonas gingivalis (P. gingivalis)* and *Tannerella forsythia (T. forsythia)*, targeting against 16S rRNA gene, have been previously reported [33,34]. PCR amplification was performed in a total reaction mixture volume of 10 µL, which included 5 µL of 2×TaqMan master mixture (LC 480 Probes Master, Roche Diagnostic GmbH), optimal concentrations of primers and hydrolysis probe (300, 300 and 200 nM for *A. actinomycetemcomitans*; 300, 300 and 300 nM for *P. gingivalis* and 300, 300 and 200 nM for *T. forsythia*) and 2.5 µL of DNA from the samples. The no-template control (NTC) consisted of 2.5 µL of sterile water. Samples were subjected to an initial amplification cycle of 95 °C for 10 min, followed by 40 cycles at 95 °C for 15 s and 60 °C for 1 min in LightCycler® 480 II thermocycler (Roche Diagnostic GmbH). Each DNA sample was analyzed in duplicate.

All assays were performed using calibration curves with a linear quantitative detection range established by the slope range of 3.3–3.6 cycles/log decade, r2 > 0.997 and an efficiency range of 1.9–2.0. Quantification was based on standard curves, which were constructed by plotting cross point cycle (Cp) values generated from qPCR against DNA extracted from serial 10-fold dilutions of purified genomic DNA from each bacterium (log of colony forming units (CFU)/mL).

2.4. Microbiological Processing by Means of Culturing

At the laboratory, aliquots of 0.1 mL from the vials were plated on two different culture media: the selective Dentaid-1 medium for the detection of *A. actinomycetemcomitans* [35] and a non-selective blood agar medium (Blood Agar Base, Oxoid, Basingstoke, England), supplemented with hemin (5 mg/L) (Sigma, St. Louis, MO, USA), menadione (1 mg/L) (Merck, Darmstadt, Germany) and 5% of sterile horse blood (Oxoid), for detection of target periodontal pathogens (*Capnocytophaga* spp., *Campylobacter rectus (C. rectus), Eikenella corrodens (E. corrodens), Fusobacterium nucleatum (F. nucleatum), P. gingivalis, Prevotella intermedia (P. intermedia), Parvimonas micra (P. micra), T. forsythia, Actinomyces odontolyticus (A. odontolyticus)*) and for evaluating total anaerobic bacterial counts. After 2–5 days of capnophilic incubation (Dentaid-1 medium) or 7–14 days of anaerobic incubation (blood agar medium), total counts and counts of representative colonies were calculated in the most suitable plates. Suspected colonies were identified by microscopy, Gram-staining and enzyme activity (Table S1). Counts were transformed in CFU per mL of the original sample.

2.5. Microbiological Processing by Means of Next Generation Sequencing

A randomly selected group of samples, from 50 subjects, 25 with periodontitis a 25 with periodontal health, were processed by means of Illumina® Sequencing (Illumina, San Diego, CA, USA) and the results have been published in a separate article [36].

2.6. Statistical Analysis

2.6.1. Sample Size Calculation

Primary outcomes were counts and frequency of detection of each target bacterial species. A sample size calculation could not be made in light of the heterogeneity of the published microbiological results. Hence, a convenience sample was selected, being larger than those reported in previous studies: 30 periodontitis patients with DS [20], 40 [19], 67 [37] or 70 [23] subjects with DS.

2.6.2. Data Analysis

Quantitative results were expressed as CFU per mL, in median values and interquartile range (IQR) and mean values and standard deviation (SD). Secondary outcomes were other microbiological variables (proportions of target pathogens) and clinical variables of the sampled sites and Ramfjord teeth, expressed as means and SD. The unit of analysis was the patient.

To assess the normality of the distribution, Shapiro-Wilk (in group with sample sizes below 30) or Kolmogorov-Smirnov test (in group with sample sizes above 30) were performed. Differences between three groups (periodontal health, gingivitis and periodontitis) were compared by one-way ANOVA test or Kruskal-Wallis test for quantitative variables and chi-square or Fisher tests for categorical variables.

The level of statistical significance was set at $p < 0.05$. A statistical software package IBMR SPSS Statistics 25.0 (IBM Corporation, Armonk, NY, USA) was used for data analysis.

3. Results

3.1. Study Population

From the 168 subjects screened, 44 were not included due to the established exclusion criteria, so the study group was composed of 124 DS subjects (Table 1 and Table S2). After evaluating their periodontal status, their distribution was: 62 with a periodontal health (50.0%), 34 with gingivitis (27.4%) and 28 with periodontitis (22.6%).

Table 1. Demographic characteristics for each study group, with the appropriate comparisons.

	Periodontal Health (H) (n = 62)						Gingivitis (G) (n = 34)						Periodontitis (p) (n = 28)						*ANOVA p* Values			
Age	mean	SD	min	max			mean	SD	min	max			mean	SD	min	max			all	H-G	G-P	H-P
age	21.4	7.5	6	40			22.6	6.0	11	34			27.9	7.2	10	42			<0.01 *	1.00	0.01 *	<0.01 *
Gender	female	male		%female	%male		female	male		%female	%male		female	male		%female	%male		all	H-G	G-P	H-P
gender	27	35		43.5%	56.5%		18	15		54.5%	45.5%		10	18		35.7%	64.3%		0.33	0.92	0.42	1.45
Systemic Conditions	none	one	≥ 2	%none	%one	% $\geq$ 2	none	one	≥ 2	%none	%one	%($\geq$2)	none	one	≥ 2	%none	%one	% $\geq$ 2	all	H-G	G-P	H-P
n systemic conditions	31	18	13	50.0%	29.0%	21.0%	13	15	6	38.2%	44.1%	17.7%	17	9	2	60.7%	32.1%	7.1%	0.52	1.51	0.53	1.48
Specific Conditions	n (no)	n (yes)		% (no)	% (yes)		n (no)	n (yes)		% (no)	% (yes)		n (no)	n (yes)		% (no)	% (yes)		all	H-G	G-P	H-P
cardiovascular diseases	62	0		100.0%	0.0%		34	0		100.0%	0.0%		27	1		96.4%	3.6%		0.22		1.35	0.93
cardiopathies	55	7		88.7%	11.3%		30	4		88.2%	11.8%		23	5		82.1%	17.9%		0.69	3.00	2.16	1.51
hypothyroidism	36	26		58.1%	41.9%		19	15		55.9%	44.1%		25	3		89.3%	10.7%		0.01 *	2.51	0.01 *	0.01 *
levothyroxine intake	37	25		59.7%	40.3%		20	14		58.8%	41.2%		25	3		89.3%	10.7%		0.01 *	2.81	0.02 *	0.01 *

SD, standard deviation; min, minimum; max, maximum; fem, female. * Statistically significant differences.

Differences in age among groups were statistically significant ($p < 0.01$), since patients with periodontitis (mean age 27.9 years) were significantly older than those with periodontal health (mean age 21.4 years; $p = 0.01$) or with gingivitis (mean age 22.6 years; $p < 0.01$). No significant differences were detected for gender, number of co-morbidities and presence of cardiovascular conditions.

Conversely, statistically significant differences among groups ($p = 0.01$) were detected for presence of hypothyroidism and intake of supplemental thyroxine medication (levothyroxine). Hypothyroidism was significantly less prevalent in periodontitis subjects (10.7%), when compared with periodontally healthy (41.9%; $p = 0.01$) and gingivitis subjects (40.3%; $p = 0.01$).

3.2. Clinical Outcome Variables

PD showed significant differences among groups for both, sampling sites and Ramfjord index teeth ($p < 0.01$), with deeper pockets in periodontitis patients (3.28 mm and 2.76 mm, respectively), than in healthy subjects (2.03 and 2.09, respectively; $p < 0.01$) and gingivitis patients (2.20 and 2.19, respectively; $p < 0.01$). Similar differences were observed for proximal PD at Ramfjord teeth and for CAL. BOP also showed significant differences among groups, for both sampling sites and Ramfjord teeth ($p < 0.01$) but differences corresponded to significant higher values in gingivitis (38.2% and 29.1%, respectively; $p < 0.01$) and periodontitis patients (50.9% and 35.8%, respectively; $p < 0.01$), when compared with healthy subjects (7.8% and 5.7%, respectively). Similar results were observed for PlI, with lower values in the periodontal health group and significantly higher levels in gingivitis and periodontitis patients, with no differences between them (Table 2 and Table S3).

3.3. Subgingival Microbiota as Evaluated by Means of qPCR

Differences in PCR counts were significantly different among groups ($p = 0.01$), with significantly higher counts in periodontitis patients, when compared with the periodontal health ($p = 0.01$) or gingivitis ($p = 0.02$) groups (Table 3 and Table S4).

Frequency of detection and counts of *A. actinomycetemcomitans* were low in all groups, although with increasing frequencies and counts in periodontitis subjects but differences were not statistically significant. *P. gingivalis*, also showed increasing frequencies and counts in periodontitis but, again, differences were not statistically significant. *T. forsythia*, showed high frequencies of detection (prevalence) in all groups (67.9% in periodontitis, 47.0% in gingivitis and 48.8% in periodontal health), with no statistically significant differences among groups, while PCR counts demonstrated significant differences ($p = 0.01$), with higher percentages in periodontitis when compared with gingivitis ($p = 0.02$) or periodontally healthy ($p = 0.01$) subjects.

3.4. Subgingival Microbiota as Evaluated by Means of Culture

Total anaerobic counts were significantly different among groups ($p < 0.01$), with significantly higher counts in periodontitis, as compared with periodontal health ($p < 0.01$).

Only four target bacterial species (*P. gingivalis*, *P. intermedia*, *T. forsythia* and *E. corrodens*) showed an overall increase in frequency of detection, counts and proportions, as the periodontal status worsened. *P. micra* and *A. odontolyticus* demonstrated higher counts, proportions and frequencies in healthy subjects. *C. rectus* and *Capnocytophaga* spp. presented the highest counts, proportions and frequencies in gingivitis patients. Finally, *A. actinomycetemcomitans* was not detected with culture methods.

Statistically significant differences among groups were only detected for *E. corrodens* ($p = 0.01$), corresponding to higher frequencies of detection ($p = 0.01$), proportions ($p = 0.01$) and counts ($p = 0.01$) in periodontitis patients, when compared with periodontal health subjects (Tables 4 and 5, Tables S5 and S6).

Table 2. Periodontal clinical outcomes, expressed as medians and interquartile ranks (IQR), for each study group, with the appropriate comparisons.

| | PERIODONTAL HEALTH (H) (n = 62) | | | | GINGIVITIS (G) (n = 34) | | | | PERIODONTITIS (P) (n = 28) | | | | *Kruskal-Wallis P Values* | | | | | | | |
| | Sampling Sites | | Ramfjord Teeth | | Sampling Sites | | Ramfjord Teeth | | Sampling Sites | | Ramfjord Teeth | | Sampling Sites | | | | Ramfjord Teeth | | | |
	Median	IQR	Median	IQR	Median	IQR	Median	IQR	Median	IQR	Median	IQR	All	H-G	G-P	H-P	all	H-G	G-P	H-P
PD (mm)	2.00	0.25	2.04	0.17	2.25	0.50	2.10	0.26	3.25	0.75	2.68	0.63	<0.01 *	0.06	<0.01 *	<0.01 *	<0.01 *	0.25	<0.01 *	<0.01 *
proximal PD (mm)			2.08	0.25			2.25	0.33			3.00	0.65					<0.01 *	0.07	<0.01 *	<0.01 *
gingival REC (mm)	0.00	0.00	0.00	0.00	0.00	0.00	0.00	0.00	0.00	0.00	0.00	0.03	0.45				0.42			
CAL (mm)	2.00	0.25	1.88	0.42	2.25	0.50	1.67	0.47	3.25	0.75	2.81	0.63	<0.01 *	0.25	<0.01 *	<0.01 *	<0.01 *	0.15	<0.01 *	<0.01 *
BOP	0.0%	0.0%	4.2%	8.3%	25.0%	25.0%	25.0%	21.1%	50.0%	50.0%	35.4%	29.2%	<0.01 *	<0.01 *	0.51	<0.01 *	<0.01 *	0.00	1.00	<0.01 *
plaque index (PlI)	0.5%	0.5%	45.8%	45.8%	75.0%	50.0%	81.3%	43.8%	75.0%	43.8%	77.5%	41.7%	<0.01 *	0.01 *	1.00	<0.01 *	<0.01 *	<0.01 *	1.00	<0.01 *
proximal PlI			58.3%	66.7%			85.4%	35.4%			83.3%	33.3%					<0.01 *	<0.01 *	1.00	<0.01 *

PD, probing depth; REC, recession; CAL: clinical attachment level; BOP, bleeding on probing. * Statistically significant differences.

Table 3. Microbiological findings (quantitative polymerase chain reaction, qPCR), with PCR counts expressed as means and standard deviations (SD) or as medians and interquartile ranks (IQR) and frequencies of detection as percentages, for the complete study population and for each study group, with the appropriate comparisons.

qPCR Counts–Mean & SD	All Subjects (*n* = 124)				
	Mean	SD	*n*	(+)	prev.
A. actinomycetemcomitans	6857	41,713	124	11	8.87%
p. gingivalis	64,064	458,060	124	15	12.10%
T. forsythia	342,644	1,755,259	124	65	52.42%

qPCR counts–mean & SD	Periodontal Health (H) (*n* = 62)			Gingivitis (G) (*n* = 34)			Periodontitis (*p*) (*n* = 28)		
	mean	SD	*n*	mean	SD	*n*	mean	SD	*n*
A. actinomycetemcomitans	406	2077	62	5601	22,125	34	22,668	83,438	24
p. gingivalis	5591	35,883	62	6396	33,297	34	263,564	947,768	23
T. forsythia	88,497	249,908	62	45,934	99,763	34	1,265,688	3,569,224	9

qPCR counts–median & IQR	Periodontal Health (H) (*n* = 62)					Gingivitis (G) (*n* = 34)					Periodontitis (*p*) (*n* = 28)					Counts (Kruskal-Wallis *p* value)				Frequency of Detection (*p* value)			
	median	IQR	*n*	(+)	prev.	median	IQR	*n*	(+)	prev.	median	IQR	*n*	(+)	prev.	all	H-G	G-*p*	H-P	all	H-G	G-P	H-P
A. actinomycetemcomitans	0	0	62	4	6.45%	0	0	34	3	8.82%	0	0	24	4	14.29%	0.41				0.51	2.09	2.07	0.75
P. gingivalis	0	0	62	6	9.68%	0	0	34	4	11.76%	0	0	23	5	17.86%	0.41				0.56	2.22	2.16	0.92
T. forsythia	0	15,625	62	30	48.39%	0	17,875	34	16	47.06%	22,350	1,372,000	9	19	67.86%	0.01 *	1.00	0.02 *	0.01 *	0.18	2.70	0.30	0.26

n, number of samples; (+), samples with pathogen detection; prev., frequency of detection of the target pathogen; A. actinomycetemcomitans: Aggregatibacter actinomycetemcomitans; P. gingivalis: Porphyromonas gingivalis; T. forsythia: Tannerella forsythia.* Statistically significant differences.

Table 4. Microbiological findings (culture), with counts (colony forming units per mL) expressed as medians and interquartile ranks (IQR) and frequencies of detection expressed as percentage, for each study group, with the appropriate comparisons.

Culture Counts	Periodontal Health (H) (n = 62)					Gingivitis (G) (n = 34)					Periodontitis (P) (n = 28)					Counts (p Values) ^		Frequency of Detection (p Value)		
	Median	IQR	n	(+)	Prev.	Median	IQR	n	(+)	Prev.	Median	IQR	n	(+)	prev.	all	all	H-G	G-P	H-P
Total anaerobic bacteria	1,435,000	1,617,500				1,735,000	2,402,500				4,065,000	4,957,500				<0.01 *				
A. actinomycetemcomitans	0	0	62	0	0.0%	0	0	34	0	0.0%	0	0	28	0	0.0%					
P. gingivalis	0	0	62	11	17.7%	0	0	34	7	20.6%	0	7505	28	8	28.6%	0.37	0.50	2.20	1.40	0.73
P. intermedia	0	625	62	22	35.5%	0	1250	34	14	41.2%	600	9000	28	15	53.6%	0.13	0.27	1.74	0.99	0.32
T. forsythia	580,645	0	62	1	1.6%	0	60,000	34	3	8.8%	0	0	28	2	7.1%	0.23	0.21	0.38	3.00	0.68
P. micra	0	0	62	4	6.5%	0	0	34	2	5.9%	0	0	28	0	0.0%	0.40	0.20	3.00	1.49	0.92
F. nucleatum	20,000	33,250	62	60	96.8%	20,000	68,125	34	29	85.3%	20,000	27,750	28	27	96.4%	0.99	0.10	0.28	0.63	3.00
C. rectus	0	0	62	3	4.8%	0	0	34	4	11.8%	0	0	28	2	7.1%	0.47	0.48	0.72	2.04	1.93
E. corrodens	0	0	62	8	12.9%	0	0	34	6	17.6%	0	10,000	28	11	39.3%	0.01 *	0.01 *	1.67	0.17	0.01 *
Capnocytophaga spp.	0	0	62	6	9.7%	0	0	34	7	20.6%	0	0	28	3	10.7%	0.25	0.32	0.63	1.47	3.00
A. odontolyticus	0	0	62	5	8.1%	0	0	34	1	2.9%	0	0	28	1	3.6%	0.54	0.49	1.25	1.47	1.98

^ Counts (Kruskal-Wallis *p* value). For comparison between groups, they were available just for anaerobic counts (H-G, 0.25; G-P, 0.43; H-P, 0.00 *) and *E. corrodens* (H-G, 1.00; G-P, 0.11; H-P, 0.01 *). * Statistically significant differences. n, number of samples; (+), samples with pathogen detection; prev., frequency of detection of the target pathogen; *A. actinomycetemcomitans*: *Aggregatibacter actinomycetemcomitans*; *P. gingivalis*: *Porphyromonas gingivalis*; *T. forsythia*: *Tannerella forsythia*; *P. micra*: *Parvimonas micra*; *F. nucleatum*: *Fusobacterium nucleatum*; *C. rectus*: *Campylobacter rectus*; *E. corrodens*: *Eikenella corrodens*; *A. odontolyticus*: *Actinomyces odontolyticus*.

Table 5. Microbiological findings (culture), with proportions of total anaerobic microbiota in percentage, expressed as medians and interquartile ranks (IQR), for each study group, with the appropriate comparisons.

	Periodontal Health (H) (*n* = 62)		Gingivitis (G) (*n* = 34)		Periodontitis (P) (*n* = 28)		*p* Value ˆ
Proportions	Median	IQR	Median	IQR	Median	IQR	All
A. actinomycetemcomitans	0.00%	0.00%	0.00%	0.00%	0.00%	0.00%	
P. gingivalis	0.00%	0.00%	0.00%	0.00%	0.00%	0.58%	0.37
P. intermedia	0.00%	0.05%	0.00%	0.06%	0.01%	0.41%	0.24
T. forsythia	0.00%	0.00%	0.00%	0.00%	0.00%	0.00%	0.24
P. micra	0.00%	0.00%	0.00%	0.00%	0.00%	0.00%	0.40
F. nucleatum	1.29%	2.77%	0.91%	3.31%	0.59%	1.76%	0.21
C. rectus	0.00%	0.00%	0.00%	0.00%	0.00%	0.00%	0.48
E. corrodens	0.00%	0.00%	0.00%	0.00%	0.00%	0.28%	0.01 *
Capnocytophaga spp.	0.00%	0.00%	0.00%	0.00%	0.00%	0.00%	0.26
A. odontolyticus	0.00%	0.00%	0.00%	0.00%	0.00%	0.00%	0.51

ˆ Proportions of total anaerobic microbiota (Kruskal-Wallis p value); for comparison between groups, they were available just for *E. corrodens* (H-G, 1.00; G-P, 0.13; H-P, 0.01 *). * Statistically significant differences. *A. actinomycetemcomitans*: *Aggregatibacter actinomycetemcomitans*; *P. gingivalis*: *Porphyromonas gingivalis*; *T. forsythia*: *Tannerella forsythia*; *P. micra*: *Parvimonas micra*; *F. nucletum*: *Fusobacterium nucleatum*; *C. rectus*: *Campylobacter rectus*; *E. corrodens*: *Eikenella corrodens*; *A. odontolyticus*: *Actinomyces odontolyticus*.

4. Discussion

In the present study, a large cohort of 124 subjects with DS showed significantly higher PCR counts of *T. forsythia*, culture counts of *E. corrodens* and total anaerobic counts, when comparing periodontitis with periodontal health or gingivitis. Subjects with periodontitis were also significantly older and the prevalence of hypothyroidism and levothyroxine intake significantly lower, as compared with the periodontal health and gingivitis groups.

Studies that have previously assessed the subgingival microbiota in DS subjects, have included a limited number of subjects and applied culture-independent techniques [19,20,23,37]. In the present study, a combination of culture and molecular techniques were selected, since they can be considered as complementary methodologies [32,38]. Using these techniques, higher counts and frequencies of detection of relevant periodontal pathogens, such as *P. gingivalis, A. actinomycetemcomitans, P. intermedia, T. forsythia* and *E. corrodens*, were identified in periodontitis, when compared with gingivitis and periodontally healthy DS subjects, as it has been previously reported in periodontitis patients without DS [39]. These quantitative and qualitative differences in the subgingival microbiota of periodontitis subjects may, in part, explain the early onset and progression of periodontitis in DS subjects, as it has been highlighted in previous studies [20,37]. However, in the present study, statistically significant differences among groups were only observed for culture counts of *E. corrodens* and total anaerobic counts, between periodontal health and periodontitis patients, and for qPCR counts of *T. forsythia*, when comparing periodontitis with periodontal health or gingivitis patients. This microbiological profile is also in agreement with the results obtained using Next Generation Sequencing (NGS), reported in a subset of this study population, which showed significantly higher levels of the genera *Tannerella, Porphyromonas* and *Aggregatibacter* in periodontitis, as compared with the periodontal health group [36].

In the present study, when using qPCR, *T. forsythia* was the most prevalent bacterial species in periodontitis patients (67.9%) but also in the other two groups. A similar trend has been reported in previous studies: Martinez-Martinez et al. [20] reported that *T. forsythia* was the most prevalent species in periodontitis DS patients (95.5%), being significantly higher when compared with periodontal health; Amano et al. [37] reported a high frequency of detection for *T. forsythia* (89.7%) in periodontitis DS patients, also significantly higher than in gingivitis patients; and *T. forsythia* was the only bacterial species with significantly higher presence and counts in DS subjects, when compared with healthy controls and individuals with cerebral palsy [23].

For *A. actinomycetemcomitans,* low frequencies of detection in the three study groups were reported in the present study. Much higher frequencies were detected in previous studies [20,37]. These differences could be explained by geographical variability, since a low prevalence of this bacterium has been reported in Spain [40,41] or due to technical aspects, such as the use of different probe-primers in the PCR technique or to differences in the patient selection process.

Higher counts and frequencies, for *T. forsythia* and *A. actinomycetemcomitans,* were reported when using qPCR, compared with culture. Similar results have been found in previous comparative studies [42]. These differences could be explained by the ability of qPCR to detect DNA of both viable and non-viable bacteria and to the lower detection limits of qPCR [32,38]. In addition, detection for *T. forsythia* by culturing is challenging [43,44], resulting in higher sensitivity and lower specificity when comparing qPCR technique *versus* culture [32,43,45]. However, the overall microbial tendency of an increase in prevalence for the three main target bacterial species from periodontal health to gingivitis and to periodontitis, was similar irrespective of the applied microbiological diagnostic technique.

The use of different microbiological technologies makes difficult to interpret the results from different studies, although common trends can be identified. In the present study, in which two complementary techniques were used, qPCR yielded higher counts for the three main target bacterial species, and higher prevalence for *T. forsythia* and *A. actinomycetemcomitans,* than culturing techniques, but the same tendency was observed with both methods, that is, an increase in prevalence/counts for the three main target bacterial species from health to gingivitis and to periodontitis. Both techniques are restricted in terms of target bacterial species, in this case three for qPCR and ten for culture. Other techniques may have a wider target such a checkerboard DNA-DNA hybridization: in the study by Sakellari et al. [23], 14 species were targeted and *T. forsythia* was identified as a relevant pathogen in all age cohorts, while *P. gingivalis, C. rectus, P. intermedia* and *A. actinomycetemcomitans* were relevant (among others) in the older cohorts; conversely, in the study by Khocht et al. [19], 40 species were targeted and *Treponema socrankii* showed significantly higher levels in DS subjects with periodontitis, as compared to individuals without periodontitis. These restrictions in the number of target species can be solved today using Next Generation Sequencing approaches, as shown in a publication based on subset of samples of the present study, but only comparing periodontitis and healthy periodontal subjects [36]: significant differences were observed at different taxonomical levels, with periodontitis patients demonstrating not only higher levels of periodontal pathogens, including *Tannerella, Treponema, Porphyromonas* and *Aggregatibacter* but also of new putative pathogens, such as species of *Peptostreptococcus, Filifactor, Fretibacterium* and *Desulfobulbus;* subjects with periodontal health showed more frequently species of *Veillonella, Neisseria, Gamella* and *Granulicatella.*

Different case definitions have been used to assess the periodontal status of DS subjects. For example, López-Pérez et al. [16] defined four categories as sound (CAL, 0 to 1 mm), mild periodontitis (CAL, 2 to 3 mm), moderate periodontitis (CAL, 4 to 5 mm) and severe periodontitis (CAL, 6 mm or more); while Khocht et al. [19] defined periodontitis as 5% or more teeth with CAL of 5 mm or more. In the present study, a more solid reference, the one proposed by the 2018 Classification of Periodontal and Peri-Implant Diseases and Conditions [17,28,29], was applied, with some modifications for the optimal use of the registered variables. The significant differences in PD, CAL and BOP among the studied categories were expected, since those variables were part of the case definition. For PlI, closely associated with BOP, it should be highlighted the large magnitude of the differences between periodontal health (around 45%), when compared with gingivitis and periodontitis (over 70%). The influence of this finding in the overall results deserve a careful analysis and the influence of a poor plaque control in the onset of gingivitis and periodontitis should be considered. In light of the possible impact of the limited skills in supragingival biofilm control in DS individuals, this factor should be taken into consideration [11].

In the present study, patients with periodontitis presented a significantly lower prevalence of hypothyroidism (10%) and the use of substitute medication (levothyroxine), compared with periodontal health and gingivitis subjects (42–44%). Hypothyroidism is the most common pathological hormone deficiency and, among the known risk factors for developing hypothyroidism, DS has been listed, usually as an autoimmune thyroiditis [46], which is consistent with the higher prevalence of autoimmune diseases in DS. Thus, the estimated prevalence of thyroid disorders in DS subjects reaches 40% in some series [47] and thanks to systematic screening, usually hypothyroidism is detected in subclinical stages in DS individuals [48], being subclinical hypothyroidism the most common detected thyroid abnormality in DS subjects [49]. Since the rate of conversion to overt hypothyroidism has been reported to be low in follow up studies [50], treatment of subclinical hypothyroidism is only advised by most authors in case of conversion to overt hypothyroidism [51].

Since the endocrine system can modulate the immune system in a bidirectional manner [52] and, in a recent scoping review, on 29 selected articles, a positive relationship between hypothyroidism and periodontitis was found [53], thus the findings of the present study are totally unexpected. One possible explanation is that periodontitis patients are actually suffering subclinical hypothyroidism and/or they have not yet been properly diagnosed [54], and, this were true, the surveillance of hypothyroid hormone function can be further supported in DS individuals. However, only speculations can be made at this point, since the previous hypothesis cannot be tested now.

In the present study, patients with periodontitis were significantly older that those in periodontal health or gingivitis groups. This finding agrees with previous reports, that have pointed out that the prevalence of periodontal diseases in DS, as it occurs in the general population, increases with age [14,15]. It is also relevant to highlight that the prevalence of periodontitis in DS individual is relatively high at a young age, ranging 58–96% in subjects younger that 35 years old [12]. As for other periodontitis as a manifestation of systemic diseases, periodontitis in DS individual presents more frequently with earlier onset, faster and generalized and more severe progression, as compared with individuals without a systemic immune compromised [16,23]. A longitudinal study has also shown that periodontal pathogens can be frequently detected in DS subjects with periodontitis, for up to 6 months, despite periodontal treatment and frequent supportive periodontal care, suggesting a higher risk of progression and giving more relevance to supragingival biofilm control as performed by the subject [55]. The importance of poor oral hygiene has been also emphasized in a recent study, suggesting a larger impact of a poor oral hygiene and an impaired immune response in the development of periodontal diseases in DS subjects [56]. However, the possible relevance of specific periodontal pathogens should not be discarded, since a recent 24-month study has even reported an association of high counts of *P. gingivalis* with risk of disease progression [57], although the study population did not include DS patients.

The present study has clinically and microbiologically evaluated a large group of DS individuals. However, some limitations have to be acknowledged, being the main limitation that periodontal diagnosis was based on a partial-mouth clinical evaluation (Ramfjord teeth), with no radiographical evaluation, which may have underestimated disease levels. This approach was selected to ease the evaluation of the special patient population included in the study and knowing that partial-mouth recording systems may present adequate degrees of accuracy [58]. Additional limitations can also be listed: (i) in a cross-sectional design, causal associations cannot be established [59]; (ii) some confounding factors may have not been properly controlled, such as diet, life-style and habits, care givers support in oral hygiene; and (iii) differences in age and thyroid disfunction may have impacted both clinical and microbiological findings.

5. Conclusions

In the present study, DS patients with periodontitis were characterized, as compared with those with periodontal health or gingivitis, by an older age, a lower frequency of thyroid dysfunction, deeper periodontal pockets and more attachment loss, higher prevalence of bleeding on probing and dental plaque accumulation and higher levels of anaerobic bacterial counts, *T. forsythia* and *E. corrodens*. It is, therefore, important, to promote actions for DS individuals, including surveillance of thyroid hormone function, improvement of oral hygiene measures and frequent evaluation of periodontal health, in order to make possible an early detection of disease (i.e., gingivitis) and, thus, providing adequate treatment or help in maintaining periodontal health, when it is still preserved.

Supplementary Materials: The following are available online at https://www.mdpi.com/2076-3417/11/2/778/s1, Supplementary Table S1. Presumptive identification of bacterial species in culture. Supplementary Table S2. Demographic characteristics for the complete study population. Supplementary Table S3. Periodontal clinical outcomes, expressed as means and standard deviations (SD), for the complete study population and for each study group. Supplementary Table S4. Microbiological findings (counts and frequencies of detection), as evaluated by means of quantitative polymerase chain reaction, expressed as means and standard deviations (SD), for the complete study population and for each study group. Supplementary Table S5. Microbiological findings, as evaluated by means of culture, with counts expressed as means and standard deviations (SD) and frequencies of detection expressed as percentage, for the complete study population and for each study group. Supplementary Table S6. Microbiological findings (proportions), as evaluated by means of culture, expressed as means and standard deviations (SD), for the complete study population and for each study group.

Author Contributions: M.C., M.J.M., A.O., M.C.S., contributed to data acquisition (microbiological analyses), statistical analyses and critically revised the manuscript; L.N., contributed to conception, data acquisition (clinical evaluations) and critically revised the manuscript; J.B., J.L., critically revised the manuscript; M.S., D.H., P.D., contributed to conception, data analysis and interpretation, critically revised the manuscript. All authors have read and agreed to the published version of the manuscript.

Funding: This project was co-funded by Xunta de Galicia under Ignicia Programme, Axencia Galega de Innovación, GAIN (12/08/2016; GRANT_NUMBER: IN855A). The participation of M.C. Sánchez in the project occurred within the activities of the Extraordinary Chair of DENTAID in Periodontal Research, University Complutense of Madrid, Spain.

Institutional Review Board Statement: The study was conducted in accordance with the Declaration of Helsinki of the World Medical Association (2008) and approved by the Research Ethics Committee of Santiago-Lugo, Spain (registration code: 2018/510).

Informed Consent Statement: Informed consent was obtained from all subjects involved (and their legal guardians, where appropriate) in the study.

Data Availability Statement: Data available on request due to restrictions. The data presented in this study are available on request from the corresponding author. The data are not publicly available due to privacy and ethical issues.

Conflicts of Interest: The authors declare no conflict of interest.

References

1. Lejeune, J.; Gautier, M.; Turpin, R. Study of somatic chromosomes from 9 mongoloid children. *C. R. Hebd. Seances Acad. Sci.* **1959**, *248*, 1721–1722. [PubMed]
2. Shin, M.; Besser, L.M.; Kucik, J.E.; Lu, C.; Siffel, C.; Correa, A.; Congenital Anomaly Multistate, P.; Survival, C. Prevalence of Down syndrome among children and adolescents in 10 regions of the United States. *Pediatrics* **2009**, *124*, 1565–1571. [CrossRef] [PubMed]
3. Asim, A.; Kumar, A.; Muthuswamy, S.; Jain, S.; Agarwal, S. Down syndrome: An insight of the disease. *J. Biomed. Sci.* **2015**, *22*, 41. [CrossRef] [PubMed]
4. Abanto, J.; Ciamponi, A.L.; Francischini, E.; Murakami, C.; de Rezende, N.P.; Gallottini, M. Medical problems and oral care of patients with Down syndrome: A literature review. *Spec. Care Dent.* **2011**, *31*, 197–203. [CrossRef] [PubMed]
5. Lagan, N.; Huggard, D.; Mc Grane, F.; Leahy, T.R.; Franklin, O.; Roche, E.; Webb, D.; O' Marcaigh, A.; Cox, D.; El-Khuffash, A.; et al. Multiorgan involvement and management in children with Down syndrome. *Acta Paediatr.* **2020**. [CrossRef]
6. Mubayrik, A.B. The Dental Needs and Treatment of Patients with Down Syndrome. *Dent. Clin. N. Am.* **2016**, *60*, 613–626. [CrossRef]
7. Scalioni, F.A.R.; Carrada, C.F.; Martins, C.C.; Ribeiro, R.A.; Paiva, S.M. Periodontal disease in patients with Down syndrome: A systematic review. *J. Am. Dent. Assoc.* **2018**, *149*, 628–639.e611. [CrossRef]

8. Bloemers, B.L.; van Bleek, G.M.; Kimpen, J.L.; Bont, L. Distinct abnormalities in the innate immune system of children with Down syndrome. *J. Pediatr.* **2010**, *156*, 804–809. [CrossRef]
9. Hennequin, M.; Allison, P.J.; Veyrune, J.L. Prevalence of oral health problems in a group of individuals with Down syndrome in France. *Dev. Med. Child. Neurol.* **2000**, *42*, 691–698. [CrossRef]
10. Moreira, M.J.; Schwertner, C.; Jardim, J.J.; Hashizume, L.N. Dental caries in individuals with Down syndrome: A systematic review. *Int. J. Paediatr. Dent.* **2016**, *26*, 3–12. [CrossRef]
11. Krishnan, C.S.; Kumari, B.N.; Sivakumar, G.; Iyer, S.P.; Ganesh, P.R. Evaluation of oral hygiene status and periodontal health in Down's syndrome subjects in comparison with normal healthy individuals. *J. Indian Acad. Dent. Spec. Res.* **2014**, *1*, 47–49. [CrossRef]
12. Morgan, J. Why is periodontal disease more prevalent and more severe in people with Down syndrome? *Spec. Care Dent.* **2007**, *27*, 196–201. [CrossRef] [PubMed]
13. Amaral Loureiro, A.C.; Oliveira Costa, F.; Eustaquio da Costa, J. The impact of periodontal disease on the quality of life of individuals with Down syndrome. *Downs Syndr. Res. Pract.* **2007**, *12*, 50–54. [CrossRef] [PubMed]
14. Nuernberg, M.A.A.; Ivanaga, C.A.; Haas, A.N.; Aranega, A.M.; Casarin, R.C.V.; Caminaga, R.M.S.; Garcia, V.G.; Theodoro, L.H. Periodontal status of individuals with Down syndrome: Sociodemographic, behavioural and family perception influence. *J. Intellect. Disabil. Res.* **2019**, *63*, 1181–1192. [CrossRef]
15. van de Wiel, B.; van Loon, M.; Reuland, W.; Bruers, J. Periodontal disease in Down's syndrome patients. A retrospective study. *Spec. Care Dent.* **2018**, *38*, 299–306. [CrossRef]
16. Lopez-Perez, R.; Borges-Yanez, S.A.; Jimenez-Garcia, G.; Maupome, G. Oral hygiene, gingivitis, and periodontitis in persons with Down syndrome. *Spec. Care Dent.* **2002**, *22*, 214–220. [CrossRef]
17. Caton, J.G.; Armitage, G.; Berglundh, T.; Chapple, I.L.C.; Jepsen, S.; Kornman, K.S.; Mealey, B.L.; Papapanou, P.N.; Sanz, M.; Tonetti, M.S. A new classification scheme for periodontal and peri-implant diseases and conditions-Introduction and key changes from the 1999 classification. *J. Clin. Periodontol.* **2018**, *45* (Suppl. 20), S1–S8. [CrossRef]
18. Jepsen, S.; Caton, J.G.; Albandar, J.M.; Bissada, N.F.; Bouchard, P.; Cortellini, P.; Demirel, K.; de Sanctis, M.; Ercoli, C.; Fan, J.; et al. Periodontal manifestations of systemic diseases and developmental and acquired conditions: Consensus report of workgroup 3 of the 2017 World Workshop on the Classification of Periodontal and Peri-Implant Diseases and Conditions. *J. Clin. Periodontol.* **2018**, *45* (Suppl. 20), S219–S229. [CrossRef]
19. Khocht, A.; Yaskell, T.; Janal, M.; Turner, B.F.; Rams, T.E.; Haffajee, A.D.; Socransky, S.S. Subgingival microbiota in adult Down syndrome periodontitis. *J. Periodontal Res.* **2012**, *47*, 500–507. [CrossRef]
20. Martinez-Martinez, R.E.; Loyola-Rodriguez, J.P.; Bonilla-Garro, S.E.; Patino-Marin, N.; Haubek, D.; Amano, A.; Poulsen, K. Characterization of periodontal biofilm in Down syndrome patients: A comparative study. *J. Clin. Pediatr. Dent.* **2013**, *37*, 289–295. [CrossRef]
21. Khocht, A.; Russell, B.; Cannon, J.G.; Turner, B.; Janal, M. Oxidative burst intensity of peripheral phagocytic cells and periodontitis in Down syndrome. *J. Periodontal Res.* **2014**, *49*, 29–35. [CrossRef] [PubMed]
22. Hajishengallis, G.; Lamont, R.J. Beyond the red complex and into more complexity: The polymicrobial synergy and dysbiosis (PSD) model of periodontal disease etiology. *Mol. Oral Microbiol.* **2012**, *27*, 409–419. [CrossRef] [PubMed]
23. Sakellari, D.; Arapostathis, K.N.; Konstantinidis, A. Periodontal conditions and subgingival microflora in Down syndrome patients. A case-control study. *J. Clin. Periodontol.* **2005**, *32*, 684–690. [CrossRef] [PubMed]
24. von Elm, E.; Altman, D.G.; Egger, M.; Pocock, S.J.; Gotzsche, P.C.; Vandenbroucke, J.P.; Initiative, S. The Strengthening the Reporting of Observational Studies in Epidemiology (STROBE) statement: Guidelines for reporting observational studies. *J. Clin. Epidemiol.* **2008**, *61*, 344–349. [CrossRef] [PubMed]
25. Ramfjord, S.P. Indices for Prevalence and Incidence of Periodontal Disease. *J. Periodontol.* **1959**, *30*, 51–59. [CrossRef]
26. O'Leary, T.J.; Drake, R.B.; Naylor, J.E. The plaque control record. *J. Periodontol.* **1972**, *43*, 38. [CrossRef] [PubMed]
27. Ainamo, J.; Bay, I. Problems and proposals for recording gingivitis and plaque. *Int. Dent. J.* **1975**, *25*, 229–235.
28. Chapple, I.L.C.; Mealey, B.L.; Van Dyke, T.E.; Bartold, P.M.; Dommisch, H.; Eickholz, P.; Geisinger, M.L.; Genco, R.J.; Glogauer, M.; Goldstein, M.; et al. Periodontal health and gingival diseases and conditions on an intact and a reduced periodontium: Consensus report of workgroup 1 of the 2017 World Workshop on the Classification of Periodontal and Peri-Implant Diseases and Conditions. *J. Clin. Periodontol.* **2018**, *45* (Suppl. 20), S68–S77. [CrossRef]
29. Tonetti, M.S.; Greenwell, H.; Kornman, K.S. Staging and grading of periodontitis: Framework and proposal of a new classification and case definition. *J. Clin. Periodontol.* **2018**, *45* (Suppl. 20), S149–S161. [CrossRef]
30. Mombelli, A.; McNabb, H.; Lang, N.P. Black-pigmenting gram-negative bacteria in periodontal disease. II. Screening strategies for detection of P. gingivalis. *J. Periodontal Res.* **1991**, *26*, 308–313. [CrossRef]
31. Syed, S.A.; Loesche, W.J. Survival of human dental plaque flora in various transport media. *Appl. Microbiol.* **1972**, *24*, 638–644. [CrossRef] [PubMed]
32. Marin, M.J.; Ambrosio, N.; O'Connor, A.; Herrera, D.; Sanz, M.; Figuero, E. Validation of a multiplex qPCR assay for detection and quantification of Aggregatibacter actinomycetemcomitans, Porphyromonas gingivalis and Tannerella forsythia in subgingival plaque samples. A comparison with anaerobic culture. *Arch. Oral Biol.* **2019**, *102*, 199–204. [CrossRef] [PubMed]

33. Boutaga, K.; van Winkelhoff, A.J.; Vandenbroucke-Grauls, C.M.; Savelkoul, P.H. Comparison of real-time PCR and culture for detection of Porphyromonas gingivalis in subgingival plaque samples. *J. Clin. Microbiol.* **2003**, *41*, 4950–4954. [CrossRef] [PubMed]
34. Boutaga, K.; van Winkelhoff, A.J.; Vandenbroucke-Grauls, C.M.; Savelkoul, P.H. Periodontal pathogens: A quantitative comparison of anaerobic culture and real-time PCR. *FEMS Immunol. Med. Microbiol.* **2005**, *45*, 191–199. [CrossRef] [PubMed]
35. Alsina, M.; Olle, E.; Frias, J. Improved, low-cost selective culture medium for Actinobacillus actinomycetemcomitans. *J. Clin. Microbiol.* **2001**, *39*, 509–513. [CrossRef] [PubMed]
36. Novoa, L.; Sanchez, M.D.C.; Blanco, J.; Limeres, J.; Cuenca, M.; Marin, M.J.; Sanz, M.; Herrera, D.; Diz, P. The Subgingival Microbiome in Patients with Down Syndrome and Periodontitis. *J. Clin. Med.* **2020**, *9*, 2482. [CrossRef]
37. Amano, A.; Kishima, T.; Akiyama, S.; Nakagawa, I.; Hamada, S.; Morisaki, I. Relationship of periodontopathic bacteria with early-onset periodontitis in Down's syndrome. *J. Periodontol.* **2001**, *72*, 368–373. [CrossRef]
38. Marin, M.J.; Ambrosio, N.; Herrera, D.; Sanz, M.; Figuero, E. Validation of a multiplex qPCR assay for the identification and quantification of Aggregatibacter actinomycetemcomitans and Porphyromonas gingivalis: In vitro and subgingival plaque samples. *Arch. Oral Biol.* **2018**, *88*, 47–53. [CrossRef]
39. Ambrosio, N.; Marin, M.J.; Laguna, E.; Herrera, D.; Sanz, M.; Figuero, E. Detection and quantification of Porphyromonas gingivalis and Aggregatibacter actinomycetemcomitans in bacteremia induced by interdental brushing in periodontally healthy and periodontitis patients. *Arch. Oral Biol.* **2019**, *98*, 213–219. [CrossRef]
40. Herrera, D.; Contreras, A.; Gamonal, J.; Oteo, A.; Jaramillo, A.; Silva, N.; Sanz, M.; Botero, J.E.; Leon, R. Subgingival microbial profiles in chronic periodontitis patients from Chile, Colombia and Spain. *J. Clin. Periodontol.* **2008**, *35*, 106–113. [CrossRef]
41. Minguez, M.; Pousa, X.; Herrera, D.; Blasi, A.; Sanchez, M.C.; Leon, R.; Sanz, M. Characterization and serotype distribution of Aggregatibacter actinomycetemcomitans isolated from a population of periodontitis patients in Spain. *Arch. Oral Biol.* **2014**, *59*, 1359–1367. [CrossRef]
42. Lau, L.; Sanz, M.; Herrera, D.; Morillo, J.M.; Martin, C.; Silva, A. Quantitative real-time polymerase chain reaction versus culture: A comparison between two methods for the detection and quantification of Actinobacillus actinomycetemcomitans, Porphyromonas gingivalis and Tannerella forsythensis in subgingival plaque samples. *J. Clin. Periodontol.* **2004**, *31*, 1061–1069. [CrossRef]
43. Bankur, P.K.; Nayak, A.; Bhat, K.; Bankur, R.; Naik, R.; Rajpoot, N. Comparison of culture and polymerase chain reaction techniques in the identification of Tannerella forsythia in periodontal health and disease, an in vitro study. *J. Indian Soc. Periodontol.* **2014**, *18*, 155–160. [CrossRef] [PubMed]
44. Paster, B.J.; Boches, S.K.; Galvin, J.L.; Ericson, R.E.; Lau, C.N.; Levanos, V.A.; Sahasrabudhe, A.; Dewhirst, F.E. Bacterial diversity in human subgingival plaque. *J. Bacteriol.* **2001**, *183*, 3770–3783. [CrossRef] [PubMed]
45. Kotsilkov, K.; Popova, C.; Boyanova, L.; Setchanova, L. Comparison of culture method and real-time PCR for detection of putative periodontopathogenic bacteria in deep periodontal pockets. *Biotechnol. Biotechnol. Equip.* **2015**, *29*, 1–7. [CrossRef]
46. Roberts, C.G.; Ladenson, P.W. Hypothyroidism. *Lancet* **2004**, *363*, 793–803. [CrossRef]
47. Amr, N.H. Thyroid Disorders in Subjects with Down Syndrome: An Update. *Acta Biomed.* **2018**, *89*, 132–139. [CrossRef]
48. Karlsson, B.; Gustafsson, J.; Hedov, G.; Ivarsson, S.A.; Anneren, G. Thyroid dysfunction in Down's syndrome: Relation to age and thyroid autoimmunity. *Arch. Dis. Child.* **1998**, *79*, 242–245. [CrossRef]
49. King, K.; O'Gorman, C.; Gallagher, S. Thyroid dysfunction in children with Down syndrome: A literature review. *Ir. J. Med. Sci.* **2014**, *183*, 1–6. [CrossRef]
50. Prasher, V.; Ninan, S.; Haque, S. Fifteen-year follow-up of thyroid status in adults with Down syndrome. *J. Intellect. Disabil. Res.* **2011**, *55*, 392–396. [CrossRef]
51. Guaraldi, F.; Rossetto Giaccherino, R.; Lanfranco, F.; Motta, G.; Gori, D.; Arvat, E.; Ghigo, E.; Giordano, R. Endocrine Autoimmunity in Down's Syndrome. *Front. Horm. Res.* **2017**, *48*, 133–146. [CrossRef] [PubMed]
52. Klein, J.R. The immune system as a regulator of thyroid hormone activity. *Exp. Biol. Med.* **2006**, *231*, 229–236. [CrossRef] [PubMed]
53. Aldulaijan, H.A.; Cohen, R.E.; Stellrecht, E.M.; Levine, M.J.; Yerke, L.M. Relationship between hypothyroidism and periodontitis: A scoping review. *Clin. Exp. Dent. Res.* **2020**, *6*, 147–157. [CrossRef] [PubMed]
54. Pinto, A.; Glick, M. Management of patients with thyroid disease: Oral health considerations. *J. Am. Dent. Assoc.* **2002**, *133*, 849–858. [CrossRef] [PubMed]
55. Sakellari, D.; Belibasakis, G.; Chadjipadelis, T.; Arapostathis, K.; Konstantinidis, A. Supragingival and subgingival microbiota of adult patients with Down's syndrome. Changes after periodontal treatment. *Oral Microbiol. Immunol.* **2001**, *16*, 376–382. [CrossRef]
56. Kosaka, M.; Senpuku, H.; Hagiwara, A.; Nomura, Y.; Hanada, N. Oral infection, periodontal disease and cytokine production in adults with Down syndrome. *Med. Res. Arch.* **2017**, *5*. Available online: https://esmed.org/MRA/mra/article/view/960 (accessed on 14 January 2021). [CrossRef]
57. Kakuta, E.; Nomura, Y.; Morozumi, T.; Nakagawa, T.; Nakamura, T.; Noguchi, K.; Yoshimura, A.; Hara, Y.; Fujise, O.; Nishimura, F.; et al. Assessing the progression of chronic periodontitis using subgingival pathogen levels: A 24-month prospective multicenter cohort study. *BMC Oral Health* **2017**, *17*, 46. [CrossRef]
58. Fernandez, E.; STROBE Group. Observational studies in epidemiology (STROBE). *Med. Clin.* **2005**, *125* (Suppl. 1), 43–48. [CrossRef]
59. Relvas, M.; Tomas, I.; Salazar, F.; Velazco, C.; Blanco, J.; Diz, P. Reliability of partial-mouth recording systems to determine periodontal status: A pilot study in an adult Portuguese population. *J. Periodontol.* **2014**, *85*, e188–e197. [CrossRef]

Article

Core Microbiota Promotes the Development of Dental Caries

Jing Chen [†], Lixin Kong [†], Xian Peng [†], Yanyan Chen, Biao Ren, Mingyun Li, Jiyao Li, Xuedong Zhou * and Lei Cheng *

State Key Laboratory of Oral Diseases, National Clinical Research Center for Oral Diseases, Department of Cariology and Endodontics West China School of Stomatology, Sichuan University, Chengdu 610041, China; chenj@stu.scu.edu.cn (J.C.); 2017224035094@stu.scu.edu.cn (L.K.); pengx@scu.edu.cn (X.P.); 2017224035129@stu.scu.edu.cn (Y.C.); renbiao@scu.edu.cn (B.R.); limingyun@scu.edu.cn (M.L.); jiyaoliscu@163.com (J.L.)
* Correspondence: zhouxd@scu.edu.cn (X.Z.); chenglei@scu.edu.cn (L.C.)
† These authors contributed equally to this work.

Abstract: A previous longitudinal study about using microbiome as a caries indicator has successfully predicted early childhood caries (ECC) in healthy individuals, but there is no evidence to verify the composition of core microbiota and its pathogenicity in vitro and in vivo. Biofilm acidogenicity, *S. mutans* count, and biofilm composition were estimated by pH evaluation, colony-forming unit, and quantitative PCR, respectively. Extracellular polysaccharide production and enamel demineralization were observed by confocal laser scanning microscopy (CLSM) and transverse microradiography (TMR), respectively. A rat caries model was established for dental caries formation in vivo, and caries lesions were quantified by Keyes Scoring. We put forward that microbiota including *Veillonella parvula*, *Fusobacterium nucleatum*, *Prevotella denticola*, and *Leptotrichia wadei* served as the predictors for ECC may be the core microbiota in ECC. This study found that the core microbiota of ECC produced limited acid, but promoted growth and acidogenic ability of *S. mutans*. Besides, core microbiota could help to promote the development of biofilms. Moreover, the core microbiota enhanced the enamel demineralization in vitro and increased cariogenic potential in vivo. These results proved that core microbiota could promote the development of dental caries and plays an important role in the development of ECC.

Keywords: early childhood caries; core microbiota; *Streptococcus mutans*; biofilms; demineralization

Citation: Chen, J.; Kong, L.; Peng, X.; Chen, Y.; Ren, B.; Li, M.; Li, J.; Zhou, X.; Cheng, L. Core Microbiota Promotes the Development of Dental Caries. *Appl. Sci.* **2021**, *11*, 3638. https://doi.org/10.3390/app11083638

Academic Editor: Yoshiaki Nomura

Received: 21 March 2021
Accepted: 14 April 2021
Published: 18 April 2021

Publisher's Note: MDPI stays neutral with regard to jurisdictional claims in published maps and institutional affiliations.

1. Introduction

Early childhood caries (ECC) is one of the most prevalent infectious diseases affecting around half of children worldwide [1,2]. ECC has been poorly controlled in many countries and has become a serious public-health problem, especially among socially vulnerable groups [3]. A better understanding of the etiology and mechanism can help with the prevention, diagnosis, treatment, and public intervention of this disease and further reduce the socioeconomic burden.

Dental caries including ECC has been considered as a multifactorial disease that is affected by the host genetic status, microorganisms, diet, and time. The etiology of dental caries is chemical-driven to microorganism-driven due to the rapid development of oral microbiology related studies. The dysbiosis of the dental plaque microbiota community could initiate this disease with the presence of fermentable carbohydrates mainly from diet. The affected plaque becomes more acidogenic and acid tolerant, which causes the decrease of local pH. Once the pH is lower than 5.5, the balance between demineralization and remineralization of dental hard tissue breaks down [4]. The continuous demineralization of dental hard tissue could eventually cause the irreversible dental carious lesions.

Preventive methods and treatments focused on traditional cariogenic bacteria like *S. mutans* and *lactobacillus* have shown some effects in clinical studies [5]. However, dental plaque, in which more than 700 different bacterial species have been identified, is one

83

of the most complex microbiota communities co-existing with the human body, and the dysbiosis of dental plaque can cause diseases just like the gut microbiota [6]. It is increasingly recognized that it is not a specific pathogen, but the interactions among different bacterial species that leads to the physiological functions and pathogenic characteristics of dental plaque [7–9]. In recent years, several studies have demonstrated the existence of core microbiome in dental caries [10,11]. The existence of a "core microbiome" was first proposed by Turnbaugh et al. [12] and referred to the organisms, genes, or functions shared by all or most individuals such as the oral cavity, nasal cavity, skin, and intestinal tract.

A previous longitudinal study about using microbiome as a dental caries indicator has successfully predicted ECC in healthy individuals, and provided evidence that some microbial populations in plaque and saliva changed acutely along with ECC onset, which may be the core microbiota of ECC [13]. This microbiota prediction model can diagnose ECC from a healthy population with 70% accuracy and predict future ECC onset with 81% accuracy. However, the researchers did not conduct studies to verify the composition of core microbiota and its pathogenicity in vivo and in vitro. Therefore, to further explore the cariogenic ability of core microbiota and its impact on previously known cariogenic *S. mutans*, our study established the multiple-species biofilm model with four represented bacteria from the highest predictors of ECC [13] including *Prevotella*, *Leptotrichia*, *Veillonella*, and *Fusobacterium*, cultured with or without *S. mutans* and compared the acid production ability, biofilm structure, EPS synthesis, demineralization ability, and in vivo cariogenic potential on animal model.

2. Materials and Methods

2.1. Bacterial Strains and Culture Conditions

The bacteria used in this study were Streptococcus mutans UA159, Fusobacterium nucleatum ATCC 25586, Veillonella parvula DSM 2008, Prevotella denticola 33-5, and Leptotrichia wadei 33-10. Streptococcus mutans, Fusobacterium nucleatum, and Veillonella parvula were commercially obtained from the American Type Culture Collection (ATCC) and the Deutsche Sammlung von Mikroorganismen und Zellkulturen (DSMZ). Prevotella denticola and Leptotrichia wadei were from the National Clinical Research Center for Oral Diseases. S. mutans and L. wadei were routinely grown at 37 °C in brain heart infusion broth (BHI; Difco, Sparks, MD, USA) [14,15], F. nucleatum and P. denticola were grown in BHI supplemented with 7.7 μM hemin (Sigma, St. Louis, MO, USA) and 1.2 μM Vitamin K1 (Sigma; St. Louis, MO, USA) [16], and V. parvula was grown in BHI containing 0.6% sodium lactate [17]. All bacterial strains were grown under anaerobic conditions (80% N_2, 10% H_2, 10% CO_2) at 37 °C [18].

2.2. Biofilm Development

In order to determine the role of the core microbiota, we established three experimental biofilm groups: "Core", "S. mutans", and "Core + S. mutans". The *S. mutans* group contained only *S. mutans*. The Core group contained *V. parvula*, *F. nucleatum*, *P. denticola*, and *L. wadei*, and the Core + *S. mutans* group contained all bacteria in the Core group and *S. mutans*.

We first adjusted all bacterial suspensions to a 2×10^8 colony-forming unit (CFU)/mL. Then, the suspension of four bacteria from the core microbiota with equal volume was mixed to generate the Core group suspension for the follow-up experiment. For the Core group, a 100 μL suspension was added onto a saliva-coated glass coverslip in a 24-well cell culture plate with 1900 μL of SHI media [19]; for the Core + *S. mutans* group, a 100 μL suspension of *S. mutans* together with a 100 μL suspension of core microbiota were added to the system with 1800 μL of SHI media; for the *S. mutans* group, a 100 μL suspension of *S. mutans* and 1900 μL of SHI media were added. By doing so, the amount of *S. mutans* equaled the total amount of core microorganisms at the beginning of the biofilm formation.

The biofilms were incubated anaerobically for 24 h, 48 h, or 72 h and 1 mL of medium in each well was renewed every 24 h. The biofilm discs were harvested at the end of the incubation time and washed out by dip-washing with phosphate-buffered saline (PBS) three times [20].

2.3. Biofilm Acidogenicity and S. Mutans Counting

For each group, pH in culture medium, used as an indicator of biofilm acidogenicity, was measured each time when the medium was changed. The medium was collected and transferred to polystyrene tubes for pH evaluation with an Orion Dual Star pH/ISE electrode (Thermo Scientific, Waltham, MA, USA) [18]. For *S. mutans* counting, the biofilms were transferred to microcentrifuge tubes containing 1 mL PBS. Serial dilutions were performed in PBS and the CFU/disc of *S. mutans* was determined by plating in triplicate on MSB plates as described in previous studies [21]. Three specimens were tested for each group.

2.4. DNA Isolation and Quantitative Analysis of Biofilm Composition

The bacterial composition of the Core + *S. mutans* group and Core group was further quantified via species-specific real-time quantitative polymerase chain reaction (qPCR) by the method described before [22]. According to the manufacturer's instructions, the TIANamp Bacterial DNA Kit (TIANGEN, Beijing, China) was used to isolate and purify the total DNA of the biofilm. We used enzymatic lysis buffer (20 mM Tris-HCl, pH 8.0; 2 mM sodium EDTA, and 1.2% Triton X-100) containing 30 mg·mL-1 lysozyme to lyse the bacteria at 37 °C for 1 h. Three independent replicates from each parameter were analyzed in triplicate using a NanoDrop ND-1000 (Thermo Scientific) and stored at −20 °C before use. For qPCR, 20 μL mixture containing 10 μL of SYBR® Premix Ex Taq (Takara, Wan Chai, Hong Kong), 1.5 μL of template, and 250 nM (each) of the forward and reverse primer were placed in each well. Primer sequences are given in Table 1. Real-time PCR was performed as follows: 95 °C for 3 min, followed by 40 cycles of 95 °C for 10 s, and 56 °C for 30 s. The mean CT value was converted into the copy number for the calculation of the percentage of each strain in the biofilm. Melting curve analysis was performed on all primer sets to ensure a single peak, which indicates primer specificity.

Table 1. Primers used in this study.

Organism	PCR Primers	Source
S. mutans	Forward:5′-TTGACGGTGTTCGTGTTGAT-3′ Reverse:5′-AAAGCGATAGGCGCAGTTTA-3′	This study
V. parvula	Forward: 5′-GTAACAAAGGTGTCGTTTCTCG-3′ Reverse: 5′-CGTAACATCTTCCGAAACTTTC-3′	[23]
F. nucleatum	Forward:5′-CAACCATTACTTTAACTCTACCATGTTCA-3′ Reverse: 5′-GTTGACTTTACAGAAGGAGATTATGTAAAAATC-3′	[24]
P. enticola	Forward:5′-GGGGATAAAGTGAGGGACGT-3′ Reverse:5′-GGCGCCTACATTTCACAACA-3′	This study
L. wadei	Forward:5′-AAGCCTGCCCTGGAAACTAT-3′ Reverse:5′-CACTTCAGCGTCAGTTACCG-3′	This study

2.5. Scanning Electron Microscopy (SEM)

For SEM analysis, the 72 h biofilms were carefully fixed with 2.5% glutaraldehyde solution for 12 h at 4 °C, then dehydrated in a series of ethanol (30, 50, 70, 80, 85, 90, 95, and 100% ethanol) and sputter-coated with gold. Specimens were examined at 10,000× magnification [25]. Three specimens were tested for each group and each sample was taken with three images.

2.6. Confocal Laser Scanning Microscopy (CLSM)

The bacteria and extracellular polysaccharides (EPS) of 72 h biofilms were labeled with SYTO 9 (Molecular Probes, Invitrogen Corp., Carlsbad, CA, USA) and Alexa Fluor 647-labeled dextran conjugate (Molecular Probes), respectively, as previously described [26].

Biofilm images were captured using a confocal laser scanning microscope (Olympus FV1000, Tokyo, Japan). The image collection gates were set to 495–515 nm for SYTO 9 and 655–690 nm for Alexa Fluor 647. Each biofilm was scanned at five selected positions, and then the confocal image series was generated by optical sectioning of each of these positions. Three-dimensional reconstruction of the biofilms and the quantification of EPS/bacteria biomass were performed with IMARIS 7.0.0 (Bitplane, Zurich, Switzerland). The EPS/bacteria ratio was calculated with ImageJ software [26,27]. Three specimens were tested for each group.

2.7. Enamel Demineralization Assessment

This experimental study was carried out on the extracted bovine teeth and measured as described previously [28,29] with some modification. Briefly, the baseline surface microhardness (SMH) was measured and bovine specimens in the range from 350 KHN to 550 KHN were collected for the demineralization investigations.

Enamel blocks (4 mm × 6 mm × 2 mm) obtained from bovine incisors were embedded in ethoxyline resin. All samples were cut again and polished by hand plane, using water-cooled silicon carbide disks (800–1000 grade paper) on both sides in parallel with a thickness range of approximately 150 nm. The enamel surface outside a 1.5 × 2.5 mm area was covered with nail polish to create the treatment window. Before the demineralization of the bovine enamel block sample, three observation points were randomly made in the above-mentioned area using a microhardness tester (Vickers indenter, 50 gf, 10 s) [30].

The blocks were randomly divided into three groups and disinfected by ethylene oxide before biofilm formation. After 72 h of biofilm incubation, the enamel blocks were collected for transverse microradiography (TMR) imaging.

The tooth slices were put into the exposure box and placed in the TMR cabinet 30 cm directly below the Cu-Kα radiation. The sample collection current was 20 mA, the voltage was 20 kV, and the collection time was 30 min [31]. Then, the image acquisition software TMR 2012 (Amsterdam, The Netherlands) was used to acquire the image of the sample [32]. The images were analyzed by using image analysis software (ImageJ; version 1.42q, Wayne Rasband, NIH, USA) [33] and customized image processing software to calculate the lesion depth (LD, μm) and the integrated mineral loss (vol%·μm, ΔZ). The lesion depth was calculated using a threshold at 90% of the mineral content of sound enamel. The integrated mineral loss was calculated by integrating the difference between the mineral content (Vol.%) in sound and demineralized enamel over the depth of the lesion [34]. Three specimens were tested for each group.

2.8. Rat Caries Model and Caries Scoring

Rat models of dental caries were established according to previous studies [35]. The experiment was performed on 20 specific pathogen-free female Wister rats. Starting from the age of 20 days, all animals were fed with a cariogenic diet (Keyes 2000) and sterilized water containing 5% *w/v* sucrose ad libitum. For the first three treatment days, the water was supplemented with chloramphenicol, ampicillin, and carbenicillin [1 g/kg]). Then, the animals were screened for *S. mutans* by an oral swab streaked on MSB agar. Furthermore, PCR was performed to detect the four bacteria from the core group.

At the age of 24 days, the rats were randomized into four groups infected for seven consecutive days with core microbiota and *S. mutans* (Group A); *S. mutans* (Group B), core microbiota (Group C), and PBS solution (Group D) (approximately 10^9 CFU/mL, 0.2 mL/rat). After three weeks, the animals were euthanized by cervical dislocation for Keyes Scoring [36]. Briefly, the teeth of each rat were stained with 0.4% ammonium salt solution for the caries red. Then, the teeth were rinsed, dried, hemisectioned, and finally observed by a stereo microscope. The depth and size of the red ammonium salt solution spread into the teeth represent the severity and impact area of the caries. Five specimens were tested for each group (n = 5).

2.9. Statistical Analysis

Data were analyzed using the SPSS software v.18.0 (International Business Machines Corp., Armonk, NY, USA). The Shapiro–Wilk (W test) was used for the normality test. Independent sample Student's *t* test was employed for the comparison of *S. mutans* counts in the biofilm of Core + *S. mutans* groups and *S. mutans* groups if normally distributed. One-way ANOVA was used for pH value, demineralization depth, mineral loss, and Keyes score analysis. Student–Newman–Keuls (SNK) test was used to identify the specific differences among groups. The significance level was set at 0.05.

3. Results

3.1. S. mutans Was Promoted by Core Microbiota and the Core + S. Mutans Biofilm Showed Increased Acidogenic Ability

To have a better consistency of the ECC plaque composition, we developed the core biofilm with an even amount of the selected four genus as well as the Core + *S. mutans* biofilm with 50% of the core microbiota and 50% of the *Streptococcus mutans*. qPCR was used for the analysis of biofilm composition. In the Core group, *V. parvula* was the most dominated species within the biofilm (69.85%, 69.87%, and 66.99%), followed by *L. wadei* (14.24%, 15.59%, and 13.77%), *P. denticola* (11.31%, 12.38%, and 17.36%), and *F. nucleatum* (2.44%, 1.53%, and 1.22%) (Figure 1a). In the Core + *S. mutans* biofilm, *S. mutans* counted for 51.16%, 54.63%, and 63.86% at 24 h, 48 h, and 72 h separately, followed by *V. parvula* (28.53%, 29.59%, and 19.96%); *L. wadei* (16.27%, 9.51%, and 4.75%); *P. denticola* (2.74%, 4.87%, and 7.45%), and *F. nucleatum* (1.30%, 1.40%, and 2.51%) (Figure 1b). In order to detect the influence of the core microbiota on the growth of *S. mutans*, we measured the count of *S. mutans* in the *S. mutans* group and in the Core + *S. mutans* group. *S. mutans* was promoted by the core microbiota and demonstrated a 3.41, 5.17, and 14.63 times increase in CFU counting at 24 h, 48 h, and 72 h compared to single *S. mutans* biofilm ($p < 0.05$) (Figure 1c,d). The pH of the culture medium measured during the experiments was taken as an indirect indicator of the biofilm acidogenicity. We found that the pH of all three groups was under 5, which can serve as the biological initiation of the enamel demineralization. The Core + *S. mutans* group showed the lowest pH at all time points ($p > 0.05$), followed by the *S. mutans* group and the Core group ($p < 0.05$) (Figure 1e).

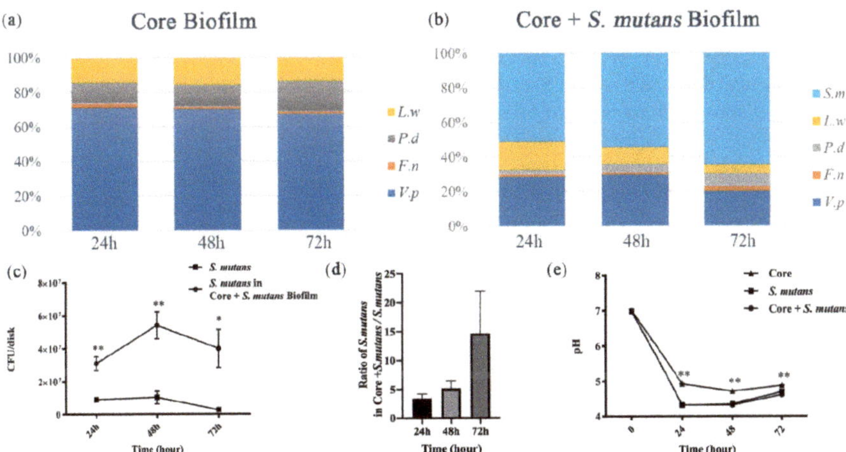

Figure 1. Composition of the biofilm and biofilm pH monitoring. (**a**) Composition of the core biofilm at 24 h, 48 h, and 72 h. (**b**) Composition of the Core + *S. mutans* biofilm at 24 h, 48 h, and 72 h. (**c**) *S. mutans* counting of the Core + *S. mutans* biofilm and *S. mutans* biofilm at 24 h, 48 h, and 72 h (n = 3). (**d**) Ratio of *S. mutans* in Core + *S. mutans*/*S. mutans*. (**e**) Medium pH of the biofilms at 24 h, 48 h, and 72 h (n = 3). Data are expressed as means ± standard errors. * $p < 0.05$; ** $p < 0.01$.

3.2. Core + S. Mutans Biofilm Showed a Net-Like Structure and Increased the Depth of the Biofilms

To have an integrated understanding of the biofilm, the SEM and CLSM were enrolled for observation of the biofilm formation, EPS synthesis, and biofilm structure in situ. The 72-h-old biofilms exhibited different morphologies under the SEM; the Core + *S. mutans* group biofilms with the existence of both EPS and bacillus represented a thick and net-like structural; the *S. mutans* group biofilms with rich EPS looked like honeycombs; the Core groups, however, without EPS, could only be detected with loose and thin biofilms (Figure 2a). Furthermore, consistent with the observation from SEM, the Core + *S. mutans* group biofilm images taken by CLSM showed a net-like structure that might improve the stability of the biofilm in complex environments (Figure 2b). The depth of the biofilms was significantly increased in the Core + *S. mutans* biofilm ($p < 0.05$) and EPS was not produced or rarely produced in the core biofilm (Figure 2c–g). The depth of the biofilm and EPS showed a significant increase in the Core + *S. mutans* group.

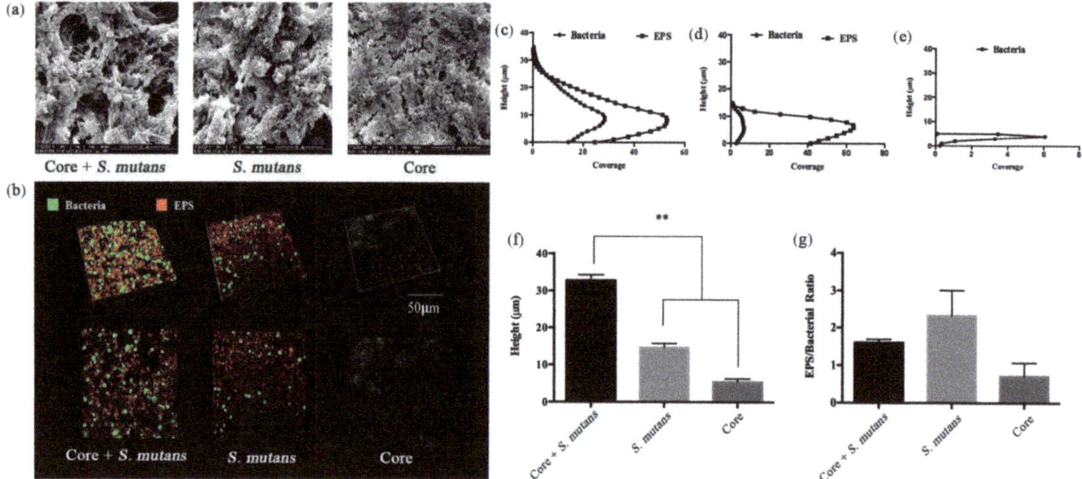

Figure 2. Biofilm structure by SEM and EPS distribution by CLSM. (**a**) Structure of the 72 h Core + *S. mutans* biofilm, *S. mutans* biofilm, and the core biofilm (n = 3). (**b**) Double-labeling of 72 h biofilms. Green, bacteria (SYTO 9); red, EPS (Alexa Fluor 647) (n = 3). (**c**) The distributions of EPS and bacteria at different heights of the Core + *S. mutans* biofilm. (**d**) The distributions of EPS and bacteria at different heights of the *S. mutans* biofilm. (**e**) The distributions of EPS and bacteria at different heights of the core biofilm. (**f**) The depth of the biofilm. (**g**) Quantification of EPS /bacteria biomass of the 72 h biofilms. ** $p < 0.01$.

3.3. Core Microbiota Enhanced the Enamel Demineralization In Vitro

To evaluate the demineralization ability of different biofilms, we harvested the biofilm demineralized enamel blocks after 72 h of treatment and tested them using transverse microradiography (TMR), which is the gold standard for detecting demineralization. The Core + *S. mutans* biofilm had the highest demineralization degree, followed by the *S. mutans* group biofilms, and the Core group had the lowest demineralization degree (Figure 3a). The integrated mineral loss was 2895 ($\pm$345.25) Vol%µm for the Core + *S. mutans* group; 1831.82 ($\pm$315.50) Vol%µm for the *S. mutans* group and 1077.78 ($\pm$175.98) Vol%µm for the Core group (Figure 3b). The demineralized depth was 72.05 ($\pm$7.18) µm for the Core + *S. mutans* group; 60.62 ($\pm$9.70) µm for the *S. mutans* group, and 53.74 ($\pm$10.43) µm for the Core group (Figure 3c). All three groups of biofilms exhibited demineralization ability on enamel and the Core + *S. mutans* biofilm had both the deepest demineralization depth and the most integrated mineral loss ($p < 0.01$). These results showed that the core microbiota promoted the enamel demineralization of *S. mutans* biofilm in vitro.

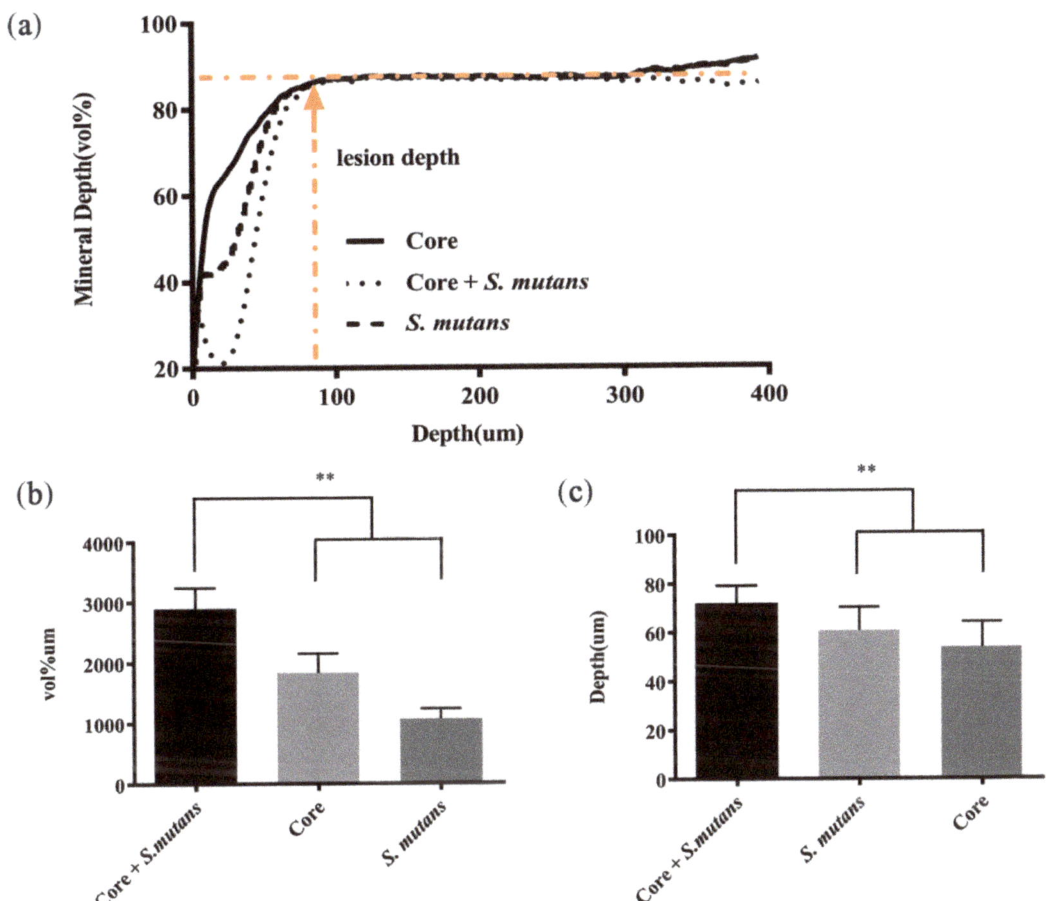

Figure 3. Enamel demineralization ability of the biofilms (n = 3). (**a**) Demineralization of the 72 h biofilm treated enamel slides. (**b**) Quantification of the integrated mineral loss. (**c**) Quantification of the demineralized depth. ** $p < 0.01$.

3.4. The Core Microbiota Increased Cariogenic Potential in a Rat Model

To explore whether the core microbiota would induce dental caries in vivo, we conducted the caries model on 20-day-old Wistar rats. For a better colonization of our selected bacteria, antibiotics were given by oral administration to eliminate the host microbiota. The bacteria were given for the first continuous seven days and the Keyes 2000 diet was fed to the rats for the whole experiment. These rats were sacrificed for Keyes scoring three weeks after bacterial challenge. As shown in Figure 4, significantly more carious lesions including enamel (E), moderate dentinal (Dm), and proximal-surface carious lesions were observed in the Core + *S. mutans* group compared with the *S. mutans* group.

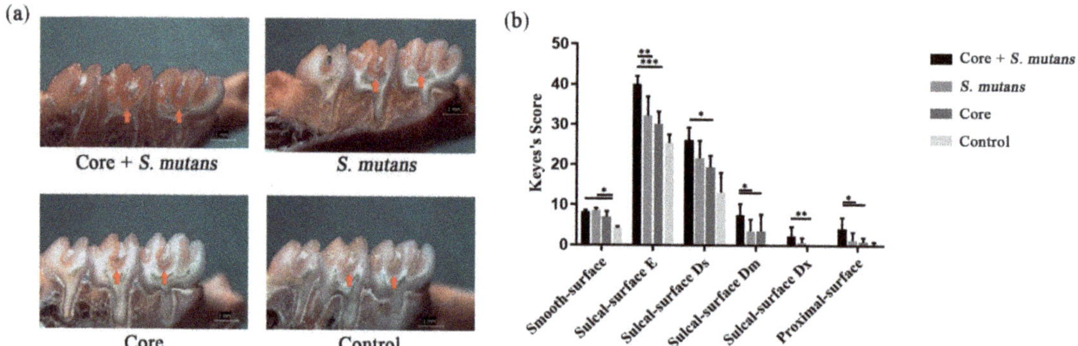

Figure 4. In vivo cariogenesis ability of different biofilms. (**a**) Representative hemisectioned molars in each group under stereoscopic microscopy. (**b**) Caries score by Keyes' method (n = 5). E = Enamel caries; Ds = slight dentinal caries; Dm = moderate dentinal caries; Dx = extensive dentinal caries. The severity decreased in the order of Core + *S. mutans*, *S. mutans*, Core and control groups. * $p < 0.05$; ** $p < 0.01$; *** $p < 0.001$.

4. Discussion

Dental caries including ECC, is a multifactorial disease [37,38]. Microbes are considered as one of the main pathogenic factors of ECC. A large amount of evidence proves that there might be a core microbiome of diseases in the oral cavity [39,40]. This also partially explains why some people are susceptible to caries while others are not. However, most of the literature have inferred the existence of core microbiome by sequencing and have found that tremendous species and genera were related to caries but failed to construct models to verify the casual relationship between dental caries and the core microbiota. A longitudinal study further narrowed the core microbiome and demonstrated that the core microbiota played an important role in ECC [13]. Referring to this literature, for the first time, we developed a core microbiota biofilm model containing four representative bacteria of ECC during its occurrence and development, which proved to promote the progress of dental caries both in vitro and in vivo.

Many studies have also shown that the four microorganisms we selected including *V. parvula*, *F. nucleatum*, *P. denticola*, and *L. wadei* are closely related to caries, especially ECC. In our study, we found that after 24 h, 48 h, and 72 h of culture, *V. parvula* accounted for more than half of the core microbial biofilm. Similarly, a large number of studies have proven that *Veillonella* content in ECC is higher than that in non-caries children, suggesting that it is closely related to ECC [13,41–44]. In another study, *Fusobacterium* was found to be significantly higher in ECC [45], in which *F. nucleatum* is closely associated with ECC, especially severe ECC [43,46]. Our study found that the amount of *F. nucleatum* in the biofilm at 72 h was about twice that at 24 h in the Core + *S. mutans* biofilm, which also seems to suggest that *F. nucleatum* is more related to severe ECC. Meanwhile, with the extension of the Core + *S. mutans* biofilm culture time, the content of *P. denticola* also increased. *P. denticola* was found to be one of the most prevalent species in the S-ECC, shown to be associated with dental caries. There is also speculation that the proportion of *P. denticola* would reduce after comprehensive restorative and preventive dental treatment [47,48]. Our results showed that when *S. mutans* was added, the content of *L. wadei* increased at the early stage of biofilm formation, which implied that *S. mutans* might promote the growth of *L. wadei*. *L. wadei* was found to be overrepresented in caries-active dental plaque compared to caries-free [49]. Moreover, a large number of *Fusobacterium*, *Prevotella*, and *Leptotrichia* were observed in a group of children with active caries [13,43,50]. In summary, these previous studies proved the rationality of our choice of four strains as core microbiota.

We found that the core microbiota had the effects of promoting the growth of *S. mutans*. There have been some controversies about the effect of *V. parvula* on *S. mutans*. Some previous studies have shown that *V. parvula* might contribute to acid production and the growth of *S. mutans* [43,51,52]. However, it has also been found that *V. parvula* had little effect on the *S. mutans* growth when co-cultured with *S. mutans* and *Streptococcus gordonii* compared with *S. mutans* alone [53]. There might be a certain relationship between *V. parvula* and *S. mutans*, but the impact of *V. parvula* in the Core microbiota on *S. mutans* remains to be studied. Similarly, *F. nucleatum* is also one of the first Gram-negative species to establish plaque biofilms [54]. It is one of the important "bridge" organisms in the naturally formed dental plaque [55], and has a central role in the ecology of dental plaque [56]. *F. nucleatum* has also been studied with the ability of affecting the growth and survival of *S. mutans*. However, the effect of core microbiota on *S. mutans* has not been previously studied. Our results showed that when the core microbiota presented, the growth of *S. mutans* increased by about four times, suggesting that the core microbiota promoted the *S. mutans* growth in the biofilms. In this study, we did not validate the function of these four species individually, but considered them as one factor. This is a limitation of our study, and we will further investigate the function of these individual species in our future studies.

The microbial composition of mature tooth biofilm is quite stable, but the pH of the biofilm could have small fluctuations [57]. A lower pH value is favored by demineralization, while a higher pH value is good for remineralization [9]. The fine-tuned balance of the oral ecosystem is disrupted, allowing disease-promoting bacteria to over grow and cause dental caries [8,37,58]. Interestingly, in our experiments, we found that the core microbiota itself produced limited acid, but could promote acid production of *S. mutans*. The integrated mineral loss of the enamel block increased by about 60% when the core microbiota existed. However, the environment in the oral cavity is very complicated, and saliva washing and food intake may have an impact on dental plaque. In vitro experiments cannot fully reproduce the complexity of in situ situation [59]. We therefore established an animal model to determine the role of the core microbiota with the presence of acidogenic and aciduric bacteria *S. mutans* and sugar. We found that rats in the Core + *S. mutans* group had the most severe dental caries.

5. Conclusions

In conclusion, we obtained core microbiota in ECC from a previous clinical research and validated its cariogenesis effects in vivo and in vitro. Subsequent clinical verification is needed in future investigations. The core microbiota containing four representative strains of ECC could promote the growth of *S. mutans*, acid production, demineralization, and the development of caries in rat (graphical abstract). This also brings new ideas and challenges to the prevention and treatment of ECC. However, how core microorganisms interact with *S. mutans* and promote the development of dental caries remain to be further studied.

Author Contributions: Conceptualization, J.C., L.K. and X.P.; Methodology, J.C., L.K., X.P. and Y.C.; Software, J.C., L.K. and X.P.; Validation, M.L. and J.L.; Formal analysis, J.C., L.K. and X.P.; Investigation, J.C., L.K. and X.P.; Resources, B.R.; Data curation, B.R., M.L. and J.L.; Writing—original draft preparation, J.C., L.K. and X.P.; Writing—review and editing, X.Z. and L.C.; Visualization, J.C., L.K. and X.P.; Supervision, X.Z. and L.C.; Project administration, X.Z. and L.C.; Funding acquisition, X.Z. and L.C. All authors have read and agreed to the published version of the manuscript.

Funding: This research was supported by the National Natural Science Foundation of China [81870759 (LC) 81430011(XZ)], the Youth Grant of the Science and Technology Department of Sichuan Province, China [2017JQ0028 (LC)], and Innovative Research Team Program of Sichuan Province (L.C). The authors declare no potential conflicts of interest with respect to the authorship and/or publication of this article.

Institutional Review Board Statement: The study was conducted according to the guidelines of the Declaration of Helsinki, and approved by the West China School of Stomatology Institute Review Board (WCSHIRB) ethics committee on 25 May 2017, and the record number was WCHSIRB-D-2017-115.

Informed Consent Statement: Not applicable.

Acknowledgments: Our thanks go to the State Key Laboratory of Oral Diseases, West China School of Stomatology of Sichuan University for their support throughout the research project.

Conflicts of Interest: The authors declare no conflict of interest.

References

1. Hemadi, A.S.; Huang, R.; Zhou, Y.; Zou, J. Salivary proteins and microbiota as biomarkers for early childhood caries risk assessment. *Int. J. Oral Sci.* **2017**, *9*, e1. [CrossRef]
2. Duangthip, D.; Chen, K.J.; Gao, S.S.; Lo, E.C.M.; Chu, C.H. Managing Early Childhood Caries with Atraumatic Restorative Treatment and Topical Silver and Fluoride Agents. *Int. J. Environ. Res. Public Health* **2017**, *14*, 1204. [CrossRef]
3. Anil, S.; Anand, P.S. Early Childhood Caries: Prevalence, Risk Factors, and Prevention. *Front. Pediatrics* **2017**, *5*, 157. [CrossRef]
4. Struzycka, I. The oral microbiome in dental caries. *Pol. J. Microbiol.* **2014**, *63*, 127–135. [CrossRef] [PubMed]
5. Hajishengallis, E.; Parsaei, Y.; Klein, M.I.; Koo, H. Advances in the microbial etiology and pathogenesis of early childhood caries. *Mol. Oral Microbiol.* **2017**, *32*, 24–34. [CrossRef]
6. Cappello, F.; Rappa, F.; Canepa, F.; Carini, F.; Mazzola, M.; Tomasello, G.; Bonaventura, G.; Giuliana, G.; Leone, A.; Saguto, D.; et al. Probiotics Can Cure Oral Aphthous-Like Ulcers in Inflammatory Bowel Disease Patients: A Review of the Literature and a Working Hypothesis. *Int. J. Mol. Sci.* **2019**, *20*, 5026. [CrossRef] [PubMed]
7. Jiang, W.; Jiang, Y.; Li, C.; Liang, J. Investigation of supragingival plaque microbiota in different caries status of Chinese preschool children by denaturing gradient gel electrophoresis. *Microb. Ecol.* **2011**, *61*, 342–352. [CrossRef] [PubMed]
8. Twetman, S. Prevention of dental caries as a non-communicable disease. *Eur. J. Oral Sci.* **2018**, *126*, 19–25. [CrossRef]
9. Manji, F.; Dahlen, G.; Fejerskov, O. Caries and Periodontitis: Contesting the Conventional Wisdom on Their Aetiology. *Caries Res.* **2018**, *52*, 548–564. [CrossRef] [PubMed]
10. Xiao, C.; Ran, S.; Huang, Z.; Liang, J. Bacterial Diversity and Community Structure of Supragingival Plaques in Adults with Dental Health or Caries Revealed by 16S Pyrosequencing. *Front. Microbiol.* **2016**, *7*, 1145. [CrossRef]
11. Yang, F.; Ning, K.; Zeng, X.; Zhou, Q.; Su, X.; Yuan, X. Characterization of saliva microbiota's functional feature based on metagenomic sequencing. *SpringerPlus* **2016**, *5*, 2098. [CrossRef] [PubMed]
12. Turnbaugh, P.J.; Ley, R.E.; Hamady, M.; Fraser-Liggett, C.M.; Knight, R.; Gordon, J.I. The human microbiome project. *Nature* **2007**, *449*, 804–810. [CrossRef] [PubMed]
13. Teng, F.; Yang, F.; Huang, S.; Bo, C.; Xu, Z.Z.; Amir, A.; Knight, R.; Ling, J.; Xu, J. Prediction of Early Childhood Caries via Spatial-Temporal Variations of Oral Microbiota. *Cell Host Microbe* **2015**, *18*, 296–306. [CrossRef] [PubMed]
14. Qiu, W.; Zheng, X.; Wei, Y.; Zhou, X.; Zhang, K.; Wang, S.; Cheng, L.; Li, Y.; Ren, B.; Xu, X.; et al. d-Alanine metabolism is essential for growth and biofilm formation of *Streptococcus mutans*. *Mol. Oral Microbiol.* **2016**, *31*, 435–444. [CrossRef] [PubMed]
15. Jang, J.Y.; Song, I.S.; Baek, K.J.; Choi, Y.; Ji, S. Immunologic characteristics of human gingival fibroblasts in response to oral bacteria. *J. Periodontal Res.* **2017**, *52*, 447–457. [CrossRef]
16. Sarkar, J.; McHardy, I.H.; Simanian, E.J.; Shi, W.; Lux, R. Transcriptional responses of *Treponema denticola* to other oral bacterial species. *PLoS ONE* **2014**, *9*, e88361. [CrossRef]
17. Poppleton, D.I.; Duchateau, M.; Hourdel, V.; Matondo, M.; Flechsler, J.; Klingl, A.; Beloin, C.; Gribaldo, S. Outer Membrane Proteome of *Veillonella parvula*: A Diderm Firmicute of the Human Microbiome. *Front. Microbiol.* **2017**, *8*, 1215. [CrossRef]
18. Chen, J.; Li, T.; Zhou, X.; Cheng, L.; Huo, Y.; Zou, J.; Li, Y. Characterization of the clustered regularly interspaced short palindromic repeats sites in *Streptococcus mutans* isolated from early childhood caries patients. *Arch. Oral Biol.* **2017**, *83*, 174–180. [CrossRef]
19. Tian, Y.; He, X.; Torralba, M.; Yooseph, S.; Nelson, K.E.; Lux, R.; McLean, J.S.; Yu, G.; Shi, W. Using DGGE profiling to develop a novel culture medium suitable for oral microbial communities. *Mol. Oral Microbiol.* **2010**, *25*, 357–367. [CrossRef]
20. Zhang, K.; Wang, S.; Zhou, X.; Xu, H.H.K.; Weir, M.D.; Ge, Y.; Li, M.; Wang, S.; Li, Y.; Xu, X.; et al. Effect of Antibacterial Dental Adhesive on Multispecies Biofilms Formation. *J. Dent. Res.* **2015**, *94*, 622–629. [CrossRef]
21. Wang, H.; Wang, S.; Cheng, L.; Jiang, Y.; Melo, M.A.S.; Weir, M.D.; Oates, T.W.; Zhou, X.; Xu, H.H.K. Novel dental composite with capability to suppress cariogenic species and promote non-cariogenic species in oral biofilms. *Mater. Sci. Eng. C Mater. Biol. Appl.* **2019**, *94*, 587–596. [CrossRef] [PubMed]
22. Ge, Y.; Ren, B.; Zhou, X.; Xu, H.H.K.; Wang, S.; Li, M.; Weir, M.D.; Feng, M.; Cheng, L. Novel Dental Adhesive with Biofilm-Regulating and Remineralization Capabilities. *Materials* **2017**, *10*, 26. [CrossRef] [PubMed]
23. Mashima, I.; Theodorea, C.F.; Thaweboon, B.; Thaweboon, S.; Nakazawa, F. Identification of *Veillonella* species in the tongue biofilm by using a novel one-step polymerase chain reaction method. *PLoS ONE* **2016**, *11*, e0157516. [CrossRef]
24. Mima, K.; Nishihara, R.; Qian, Z.R.; Cao, Y.; Sukawa, Y.; Nowak, J.A.; Yang, J.; Dou, R.; Masugi, Y.; Song, M. *Fusobacterium nucleatum* in colorectal carcinoma tissue and patient prognosis. *Gut* **2016**, *65*, 1973–1980. [CrossRef] [PubMed]
25. Cheng, L.; Zhang, K.; Weir, M.D.; Liu, H.B.; Zhou, X.D.; Xu, H.H.K. Effects of antibacterial primers with quaternary ammonium and nano-silver on *Streptococcus mutans* impregnated in human dentin blocks. *Dent. Mater.* **2013**, *29*, 462–472. [CrossRef]
26. Xiao, J.; Klein, M.I.; Falsetta, M.L.; Lu, B.W.; Delahunty, C.M.; Yates, J.R.; Heydorn, A.; Koo, H. The Exopolysaccharide Matrix Modulates the Interaction between 3D Architecture and Virulence of a Mixed-Species Oral Biofilm. *PLoS Pathog.* **2012**, *8*, e1002623. [CrossRef]
27. Heydorn, A.; Nielsen, A.T.; Hentzer, M.; Sternberg, C.; Givskov, M.; Ersbøll, B.K.; Molin, S. Quantification of biofilm structures by the novel computer program COMSTAT. *Microbiology* **2000**, *146*, 2395–2407. [CrossRef]

28. Huang, X.; Cheng, L.; Exterkate, R.A.; Liu, M.; Zhou, X.; Li, J.; ten Cate, J.M. Effect of pH on Galla chinensis extract's stability and anti-caries properties in vitro. *Arch. Oral Biol.* **2012**, *57*, 1093–1099. [CrossRef]

29. Han, Q.; Li, B.; Zhou, X.; Ge, Y.; Wang, S.; Li, M.; Ren, B.; Wang, H.; Zhang, K.; Xu, H.H.K.; et al. Anti-Caries Effects of Dental Adhesives Containing Quaternary Ammonium Methacrylates with Different Chain Lengths. *Materials* **2017**, *10*, 643. [CrossRef]

30. Beighton, D.; Gilbert, S.C.; Clark, D.; Mantzourani, M.; Al-Haboubi, M.; Ali, F.; Ransome, E.; Hodson, N.; Fenlon, M.; Zoitopoulos, L.; et al. Isolation and identification of bifidobacteriaceae from human saliva. *Appl. Environ. Microbiol.* **2008**, *74*, 6457–6460. [CrossRef]

31. Schwendicke, F.; Korte, F.; Dorfer, C.E.; Kneist, S.; Fawzy El-Sayed, K.; Paris, S. Inhibition of *Streptococcus mutans* Growth and Biofilm Formation by Probiotics in vitro. *Caries Res.* **2017**, *51*, 87–95. [CrossRef]

32. Caglar, E.; Sandalli, N.; Twetman, S.; Kavaloglu, S.; Ergeneli, S.; Selvi, S. Effect of yogurt with *Bifidobacterium* DN-173 010 on salivary mutans streptococci and lactobacilli in young adults. *Acta Odontol. Scand.* **2005**, *63*, 317–320. [CrossRef]

33. Nam, H.J.; Kim, Y.M.; Kwon, Y.H.; Kim, I.R.; Park, B.S.; Son, W.S.; Lee, S.M.; Kim, Y.I. Enamel Surface Remineralization Effect by Fluorinated Graphite and Bioactive Glass-Containing Orthodontic Bonding Resin. *Materials* **2019**, *12*, 1308. [CrossRef] [PubMed]

34. Kielbassa, A.M. In situ induced demineralization in irradiated and non-irradiated human dentin. *Eur. J. Oral Sci.* **2000**, *108*, 214–221. [CrossRef]

35. Liu, S.Y.; Wu, T.M.; Zhou, X.D.; Zhang, B.; Huo, S.B.; Yang, Y.T.; Zhang, K.K.; Cheng, L.; Xu, X.; Li, M.Y. Nicotine is a risk factor for dental caries: An *in vivo* study. *J. Dent. Sci.* **2018**, *13*, 30–36. [CrossRef] [PubMed]

36. Keyes, P.H. Dental caries in the molar teeth of rats. II. A method for diagnosing and scoring several types of lesions simultaneously. *J. Dent. Res.* **1958**, *37*, 1088–1099. [CrossRef] [PubMed]

37. Philip, N.; Suneja, B.; Walsh, L.J. Ecological Approaches to Dental Caries Prevention: Paradigm Shift or Shibboleth? *Caries Res.* **2018**, *52*, 153–165. [CrossRef] [PubMed]

38. Sanz, M.; Beighton, D.; Curtis, M.A.; Cury, J.A.; Dige, I.; Dommisch, H.; Ellwood, R.; Giacaman, R.A.; Herrera, D.; Herzberg, M.C.; et al. Role of microbial biofilms in the maintenance of oral health and in the development of dental caries and periodontal diseases. Consensus report of group 1 of the Joint EFP/ORCA workshop on the boundaries between caries and periodontal disease. *J. Clin. Periodontol.* **2017**, *44* (Suppl. 18), S5–S11. [CrossRef]

39. Jiang, W.; Zhang, J.; Chen, H. Pyrosequencing analysis of oral microbiota in children with severe early childhood dental caries. *Curr. Microbiol.* **2013**, *67*, 537–542. [CrossRef]

40. Jiang, W.; Ling, Z.X.; Lin, X.L.; Chen, Y.D.; Zhang, J.; Yu, J.J.; Xiang, C.; Chen, H. Pyrosequencing Analysis of Oral Microbiota Shifting in Various Caries States in Childhood. *Microb. Ecol.* **2014**, *67*, 962–969. [CrossRef]

41. Kanasi, E.; Dewhirst, F.E.; Chalmers, N.I.; Kent, R., Jr.; Moore, A.; Hughes, C.V.; Pradhan, N.; Loo, C.Y.; Tanner, A.C. Clonal analysis of the microbiota of severe early childhood caries. *Caries Res.* **2010**, *44*, 485–497. [CrossRef] [PubMed]

42. Ling, Z.; Kong, J.; Jia, P.; Wei, C.; Wang, Y.; Pan, Z.; Huang, W.; Li, L.; Chen, H.; Xiang, C. Analysis of oral microbiota in children with dental caries by PCR-DGGE and barcoded pyrosequencing. *Microb. Ecol.* **2010**, *60*, 677–690. [CrossRef]

43. Tanner, A.C.; Mathney, J.M.; Kent, R.L.; Chalmers, N.I.; Hughes, C.V.; Loo, C.Y.; Pradhan, N.; Kanasi, E.; Hwang, J.; Dahlan, M.A.; et al. Cultivable anaerobic microbiota of severe early childhood caries. *J. Clin. Microbiol.* **2011**, *49*, 1464–1474. [CrossRef]

44. Agnello, M.; Marques, J.; Cen, L.; Mittermuller, B.; Huang, A.; Chaichanasakul Tran, N.; Shi, W.; He, X.; Schroth, R.J. Microbiome Associated with Severe Caries in Canadian First Nations Children. *J. Dent. Res.* **2017**, *96*, 1378–1385. [CrossRef] [PubMed]

45. Corby, P.M.; Lyons-Weiler, J.; Bretz, W.A.; Hart, T.C.; Aas, J.A.; Boumenna, T.; Goss, J.; Corby, A.L.; Junior, H.M.; Weyant, R.J.; et al. Microbial risk indicators of early childhood caries. *J. Clin. Microbiol.* **2005**, *43*, 5753–5759. [CrossRef]

46. Unsal, G.; Topcuoglu, N.; Ulukapi, I.; Kulekci, G.; Aktoren, O. Scardovia Wiggsiae and the Other Microorganisms in Severe Early Childhood Caries. *J. Dent. Oral Care Med.* **2017**, *3*, 2454–3276.

47. Kamalendran, N. Oral Microbiota of Children Undergoing Comprehensive Treatment for Severe Early Childhood Caries. Ph.D. Thesis, University of Otago, Dunedin, New Zealand, 2015.

48. Hurley, E.; Barrett, M.P.J.; Kinirons, M.; Whelton, H.; Ryan, C.A.; Stanton, C.; Harris, H.M.B.; O'Toole, P.W. Comparison of the salivary and dentinal microbiome of children with severe-early childhood caries to the salivary microbiome of caries-free children. *BMC Oral Health* **2019**, *19*, 13. [CrossRef]

49. Peterson, S.N.; Snesrud, E.; Liu, J.; Ong, A.C.; Kilian, M.; Schork, N.J.; Bretz, W. The dental plaque microbiome in health and disease. *PLoS ONE* **2013**, *8*, e58487. [CrossRef] [PubMed]

50. Luo, A.H.; Yang, D.Q.; Xin, B.C.; Paster, B.J.; Qin, J. Microbial profiles in saliva from children with and without caries in mixed dentition. *Oral Dis.* **2012**, *18*, 595–601. [CrossRef]

51. Marsh, P.D. Microbial ecology of dental plaque and its significance in health and disease. *Adv. Dent. Res.* **1994**, *8*, 263–271. [CrossRef] [PubMed]

52. Gross, E.L.; Beall, C.J.; Kutsch, S.R.; Firestone, N.D.; Leys, E.J.; Griffen, A.L. Beyond *Streptococcus mutans*: Dental caries onset linked to multiple species by 16S rRNA community analysis. *PLoS ONE* **2012**, *7*, e47722. [CrossRef] [PubMed]

53. Liu, J.; Wu, C.; Huang, I.H.; Merritt, J.; Qi, F. Differential response of *Streptococcus mutans* towards friend and foe in mixed-species cultures. *Microbiology* **2011**, *157*, 2433–2444. [CrossRef] [PubMed]

54. Könönen, E. Development of oral bacterial flora in young children. *Ann. Med.* **2000**, *32*, 107–112. [CrossRef] [PubMed]

55. Kolenbrander, P.E.; Palmer, R.J., Jr.; Periasamy, S.; Jakubovics, N.S. Oral multispecies biofilm development and the key role of cell-cell distance. *Nat. Rev. Microbiol.* **2010**, *8*, 471–480. [CrossRef] [PubMed]

56. Haake, S.K.; Yoder, S.C.; Attarian, G.; Podkaminer, K. Native plasmids of *Fusobacterium nucleatum*: Characterization and use in development of genetic systems. *J. Bacteriol.* **2000**, *182*, 1176–1180. [CrossRef] [PubMed]
57. Fejerskov, O.; Scheie, A.A.; Manji, F. The Effect of Sucrose on Plaque Ph in the Primary and Permanent Dentition of Caries-Inactive and Caries-Active Kenyan Children. *J. Dent. Res.* **1992**, *71*, 25–31. [CrossRef]
58. Kilian, M.; Chapple, I.L.; Hannig, M.; Marsh, P.D.; Meuric, V.; Pedersen, A.M.; Tonetti, M.S.; Wade, W.G.; Zaura, E. The oral microbiome—an update for oral healthcare professionals. *Br. Dent. J.* **2016**, *221*, 657–666. [CrossRef] [PubMed]
59. Kantarci, A.; Hasturk, H.; Van Dyke, T.E. Animal models for periodontal regeneration and peri-implant responses. *Periodontol. 2000* **2015**, *68*, 66–82. [CrossRef]

applied sciences

MDPI

Case Report

Complications of Teeth Affected by Molar-Incisor Malformation and Pathogenesis According to Microbiome Analysis

Hyo-Seol Lee [1,2,*], Hee Jin Kim [1], Koeun Lee [2], Mi Sun Kim [1,3], Ok Hyung Nam [1,2] and Sung-Chul Choi [1,2]

[1] Department of Pediatric Dentistry, School of Dentistry, Kyung Hee University, Seoul 02447, Korea; khupedo@khu.ac.kr (H.J.K.); pedokms@khu.ac.kr (M.S.K.); pedokhyung@khu.ac.kr (O.H.N.); pedochoi@khu.ac.kr (S.-C.C.)
[2] Department of Pediatric Dentistry, Kyung Hee University Dental Hospital, Seoul 02447, Korea; olivedlr@naver.com
[3] Department of Pediatric Dentistry, Kyung Hee University Dental Hospital at Gangdong, Seoul 05278, Korea
* Correspondence: stberryfield@gmail.com or snowlee@khu.ac.kr; Tel.: +82-10-4926-2808; Fax: +82-2-965-7247

Abstract: A molar-incisor malformation (MIM) is a recently reported dental anomaly that causes premature loss of the first molar with severe dentoalveolar infection. However, there has been no research on the pathogenesis yet. The aim of this study was to report the clinical process of MIMs and investigate the pathogenesis by conducting a microbiome analysis. An eight-year-old girl was diagnosed with MIM and after two years, four permanent first molars were sequentially extracted due to severe dentoalveolar infection. We recorded the patient's clinical progress and collected oral microbiome samples from the extracted teeth with MIM and sound teeth as controls. The sites of microbiome sampling were represented by five habitats in two groups. Group (1) was the perio group: ① supragingival plaque, ② subgingival plaque, and ③ a pical abscess; and group (2) was the endo group: ④ coronal pulp chamber and ⑤ root canal. The perio group was composed predominantly of genera *Streptococcus*, *Veilonella*, and *Leptotrichia*. *Spirochetes* appeared in one sample from a severe periodontal abscess. *Aggregatibacter actinomyces* were not identified. In the endo groups, pulp necrosis was observed in all MIM and the genera *Peptostreptococcus* and *Parvimonas* predominated. In conclusion, MIM teeth caused localized tooth-related periodontitis with pulp necrosis rather than localized juvenile periodontitis, resulting in a poor prognosis, and timely extraction is highly recommended.

Keywords: Molar-incisor malformation; complication; microbiome; pathogenesis; localized tooth-related periodontitis

Citation: Lee, H.; Kim, H.J.; Lee, K.; Kim, M.S.; Nam, O.H.; Choi, S. Complications of Teeth Affected by Molar-Incisor Malformation and Pathogenesis According to Microbiome Analysis. *Appl. Sci.* **2021**, *11*, 4. https://dx.doi.org/10.3390/app11010004

Received: 13 October 2020
Accepted: 18 December 2020
Published: 22 December 2020

Publisher's Note: MDPI stays neutral with regard to jurisdictional claims in published maps and institutional affiliations.

1. Introduction

A molar-incisor malformation (MIM) is a recently reported dental anomaly of the permanent first molars, deciduous molars, and permanent maxillary central incisors [1]. MIM anomalies of the permanent first molars and deciduous molars may be characterized by normal crowns with a constricted cervical region and thin, narrow, and short roots, while the affected maxillary central incisors may exhibit a hypoplastic enamel notch near the cervical third of the clinical crown [2]. Although the etiology of MIMs remains uncertain, it is thought to be attributable to systemic disease associated with the neural system during infancy [3–6].

MIM was named by Lee et al. [1] in 2014. In addition, there have been papers published this dental anomaly as 'Root malformation associated with a cervical mineralized diaphragm' or 'Molar root-incisor malformation'. According to Vargo et al. [3] in 2020, a total 87 cases have been published in articles so far, and all except one case have been affected by the permanent first molars. The average diagnostic age was nine years.

MIM has been mistaken for molar-incisor hypomineralization (MIH) until recently. MIM has similarities with and differences from MIH. The similarities include the involved teeth; both MIM and MIH appear most frequently in the maxillary central incisors and the permanent first molars [1,2]. There are differences, however, in the affected tissues. In MIH, the enamel is affected, whereas in MIM, dentin and cementum are affected. In incisors affected by MIM, the cervical enamel notch may appear. The exact cause of both is unknown, and treatment is difficult [3]. Since it has been the biggest problem in children's oral health recently, it is important to identify the correct treatment.

MIM teeth are associated with complications such as dentoalveolar infections, early exfoliation, space loss, spontaneous pain, impaction, and poor incisor esthetics [3]. Dentoalveolar infection is the most common complication of MIM. In severe cases, inflammation surrounds the entire roots, resulting in an abscess, fistula, and vertical mobility. A MIM with complications is usually extracted due to the poor prognosis reported in previous case studies [1,7,8]. The early loss of the first permanent molar can negatively affect occlusal stability and craniofacial development [9,10].

Various factors can affect the development of a dentoalveolar infection, and the oral microbiome is one of them. The oral microbiome is described as a group of resident oral microorganisms in the host. Oral diseases, such as dental caries and periodontal disease are caused by dysbiosis of the host and oral microbiome [11]. Next-generation sequencing (NGS) is an effective method used to conduct research on the microbiome [12]. NGS is a molecular technique that reads more sequences faster and more economically and is widely used to identify microorganisms.

In this study, we analyzed the oral microbiomes around the MIM teeth to provide evidence for the pathogenesis of the complications because the oral microbiome is associated with oral health and oral disease. The aim of this study was to report the clinical process of MIM teeth and investigate the pathogenesis by microbiome analysis.

2. Materials and Methods

2.1. Ethics

The guidelines in the Helsinki Declaration were followed in this investigation. This study was approved by the Institutional Review Board of the School of Dentistry, Kyung Hee University (KHDIRB1606-6). The participants were recruited from patients in the Department of Pediatric Dentistry, School of Dentistry, Kyung Hee University. Written consent for participation in the study was obtained from the parent and child.

2.2. Clinical Progress

An eight-year-old girl visited our dental clinic for a dental eruption problem. Her medical history included a premature birth at nine months and she was taking medication for attention deficit hyperactivity disorder (ADHD). Radiographic and clinical examinations revealed root deformation of the first molars, primary molars, and cervical notch of the maxillary central incisors (Figure 1). The maxillary first molars were locked on the distal surface of the maxillary primary second molar and the mandibular first molars were mesially displaced due to the early exfoliation of the mandibular primary second molars. A lingual arch was set and periodic checks followed.

When the patient was 10, severe dentoalveolar infection and fistula occurred on the mandibular left first molar (Figure 2). Orthodontic analysis was conducted for treatment planning. She already had a space discrepancy and we decided to extract all permanent first molars affected by MIM and perform orthodontic treatment after the eruption of the secondary molars. Thereafter, the four MIM first molars were extracted sequentially and they showed twisted roots with granulation tissue (Figure 3). The crown and root of the maxillary left molar and mandibular right molar were fractured during the extraction. One year later, the second molars erupted and were relatively well-aligned.

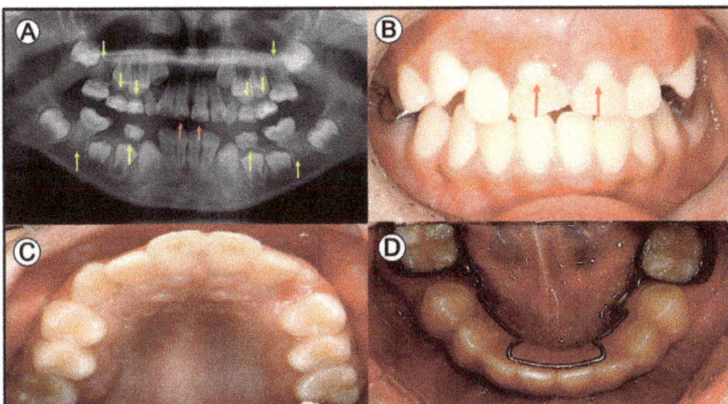

Figure 1. Panoramic radiographic and clinical photos at the first visit. (**A**) In the 8-year-old girl, all the primary molars, first molars (yellow arrows), and anterior teeth (red arrows) were affected by MIM. There was a loss of space due to early exfoliation of the mandibular primary second molars. (**B**–**D**) A mandibular lingual arch was set for space maintenance. Morphological abnormalities (red arrows) were observed in the maxillary central incisors.

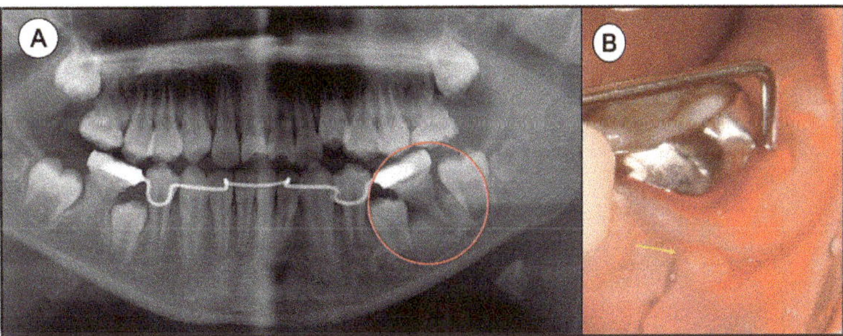

Figure 2. Clinical photos. (**A**) Two years later, a severe dentoalveolar abscess was observed on the mandibular left first molar. (**B**) The full depth of the distobuccal region was probed and a fistula was seen (yellow arrow).

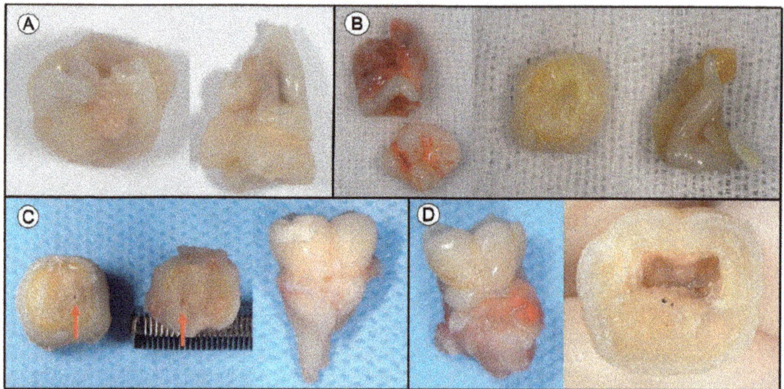

Figure 3. Extracted teeth affected with MIM. (A) The maxillary right first molar (ID 16) showed twisted and fused roots after cleaning. **(B)** The maxillary left first molar (ID 26) showed a fractured crown and root with granulation tissue and an abnormal calcified surface between the crown and root. **(C)** The mandibular right first molar (ID 46) showed a small canal (red arrow) on the fractured surface. **(D)** The mandibular left first molar (ID 36) showed abundant granulation tissue and pulp necrosis in the pulp chamber.

2.3. Sampling

The samples were MIM teeth and the non-MIM teeth were the controls. The MIM samples included four permanent first molars and maxillary central incisors as one sample, and the non-MIM teeth including the maxillary left premolars and mandibular right premolars as one sample.

The oral microbiome was obtained from five habitats around the teeth in the two groups. The perio group included ① a supragingival plaque, ② subgingival plaque, and ③ an apical abscess, and the endo group included ④ a coronal pulp chamber and ⑤ a root canal (Figure 4). The perio samples were collected before the extractions, and the endo samples were collected after the extractions. All samples are listed in Table 1. The sample identification (ID) numbers refer to the two-digit FDI notations.

- **Perio group**
    - ①   Supragingival plaque: The patient was requested not to brush her mouth 24 h prior to sample collection. Supragingival dental plaque samples were collected by rubbing a sterile cotton swab across the cervicobuccal area of the teeth twice or three times under pressure [13].
    - ②   Subgingival plaque: The subgingival plaque samples were collected by inserting sterile endodontic paper points (sized 30; two paper points per site) into the gingival sulci or periodontal pocket for 10 s right after isolation and supragingival plaque removal [14].
    - ③   Periapical abscess: The periapical abscess sample was obtained via aspiration. The swollen area was aspirated with a syringe fitted with a 16-gauge needle and expressed into a sterile vial [15].

- **Endo group**
    - ①   Coronal pulp chamber and ⑤ root canal: The extracted tooth was cleaned with 30% hydrogen peroxide [16]. Aseptic techniques such as sterile burs were used to access the pulp space. Bacteriological samples of the pulp chamber were collected immediately after crown access. Pulp remnant and infected dentin were collected using a sterilized spoon excavator and sterilized paper points were inserted to absorb the remaining fluid containing microorganisms. In the root canal, two sequential new paper points were placed at the same level and utilized to soak up the fluid in the canal. Each paper point was held in position

for 30 s. Sterilized endodontic files were then used to collect the infected root canal dentin severally. Only the tip area was collected into the tube by cutting.

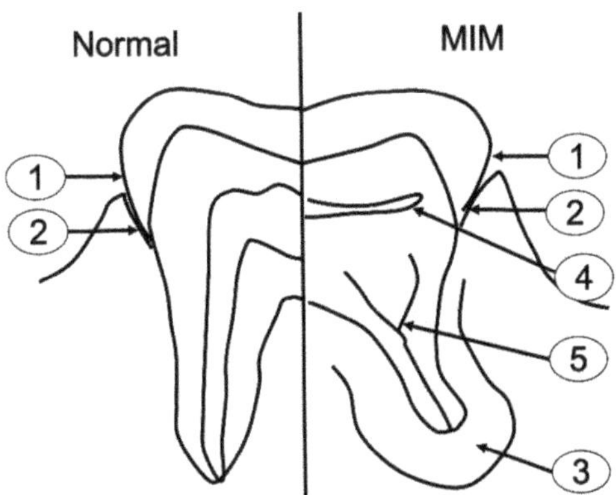

Figure 4. Sampling of normal and MIM teeth. Perio group: ① supragingival plaque, ② subgingival plaque, and ③ apical abscess. Endo group: ④coronal pulp chamber and ⑤ root canal.

Table 1. Sample identification (ID).

Group		Location	Maxillary Central Incisors	Maxillary Right First Molar	Maxillary Left First Molar	Mandibular Left First Molar	Mandibular Right First Molar	Non-MIM Teeth
			11	16	26	36	46	Nor
Perio	1	Supragingival plaque	11-1	16-1	26-1	36-1	46-1	Nor-1
	2	Subgingival plaque	11-2	16-2	26-2	36-2	46-2	Nor-2
	3	Apical abscess					46-3	
Endo	4	Coronal pulp chamber		16-4	*	36-4	**	
	5	Root canal		16-5	*	36-5	**	

* The crown and root were separated during the extraction; ** Samples were taken but DNA extraction failed due to DNA degradation or insufficient amount.

The samples were placed in sterile 1.5 mL microcentrifuge tubes and frozen for storage at −80 °C. DNA was extracted from the clinical samples using the FastDNA SPIN Kit for Soil (MP Biomedicals, Santa Ana, CA, USA) according to the manufacturer's instructions. The concentration and purity of the DNA samples were determined using an ND-1000 NanoDrop spectrophotometer (Thermo Scientific, Waltham, MA, USA) by measuring the absorbance at 260 and 280 nm.

2.4. PCR Amplification and Illumina Sequencing

PCR amplification of the extracted DNA was performed using primers targeting regions between V3 and V4 of the 16S rRNA gene. For bacterial amplification, 341F primers (5′TCGTCGGCAGCGTC-AGATGTGTATAAGAGACAG-CCTACGGGNGGCWGCAG-3′; the target region primer) and 805R (5′-GTCTCGTGGGCTCGGAGATGTGTATAAGAGACA G-GACTACHVGGGTATCTAATCC-3) were used. Amplifications were conducted under the following conditions: initial denaturation at 95 °C for 3 min, followed by 25 cycles of denaturation at 95 °C for 30 s, primer annealing at 55 °C for 30 sec, and extension at 72 °C for 30 s, with a final elongation at 72 °C for 5 min. Secondary amplification for attaching the Illumina Nextera barcode was then performed using the i5 forward primer (5′AATGATACGGCGACCACCGAGATCTACAC-XXXXXXXX-TCGTCGGCAGCGTC-3′; X indicates the barcode region) and i7 reverse primer (5′-CAAGCAGAAGACGGCATACGAG

AT-XXXXXXXXAGTCTCGTGGGCTCGG-3). The conditions for the secondary amplification were similar to the earlier one except that the amplification cycles were set to 8.

The PCR products were confirmed using 2% agarose gel electrophoresis and visualized by the Gel Doc system (BioRad, Hercules, CA, USA). The amplified products were purified via the QIAquick PCR purification kit (Qiagen, Valencia, CA, USA). Equal concentrations of purified products were pooled together and short fragments (non-target products) were removed with the Ampure bead kit (Agencourt Bioscience, MA, USA). The quality and size of the products were assessed on a Bioanalyzer 2100 (Agilent, Palo Alto, CA, USA) using a DNA 7500 chip. Mixed amplicons were pooled, and sequencing was conducted at Chunlab, Inc. (Seoul, Korea) using the Illumina MiSeq Sequencing System (Illumina, San Diego, CA, USA) according to the manufacturer's instructions.

2.5. MiSeq Pipeline Method

The processing of the raw reads started with a quality check and filtering of the low quality (<Q25) reads using Trimmomatic 0.32. After passing the QC check, the paired-end sequence data were merged using PandaSeq. The primers were then trimmed with ChunLab's in-house program at a similarity cutoff of 0.8. The noise was removed from the sequences using Mothur's pre-clustering program, which merges the sequences and extracts unique sequences allowing up to two differences between the sequences. The Ez-Taxon database was used for taxonomic assignments using BLAST 2.2.22 and pairwise alignment was used to calculate similarity. Uchime and the non-chimeric 16S rRNA database from EzTaxon were used to detect chimerism in the reads with best hit similarity rates below 97%. Sequence data were then clustered using CD-Hit and UCLUST, and alpha diversity analysis was conducted.

3. Results

3.1. Clinical Information

The left MIM (ID 26 and 36) had a severe dentoalveolar abscess with a fistula and was extracted with abundant granulation tissue. The right MIM (ID 16 and 46) showed relatively less granulation tissue. The MIM had short, twisted, and fused abnormal roots (Figure 3). ID 26 and 46 were fractured during the extraction. The fractured surfaces were roughly calcified and there was a small hole connecting the pulp chamber and the root canal. The coronal pulp chamber and root canal were opened after extraction and necrotized, liquefied, gray, and slurried pulp and infected dentin were revealed. The incisors affected by the MIM (ID 11) and the non-MIM teeth (ID Nor) were clinically sound.

3.2. Taxonomic Identification and Operational Taxonomical Unit (OTU) Assessment of Diversity

An average of 18,621, 19,051, 19,952, 18,136, and 17,402 sequence reads was generated from the supragingival, subgingival, apical abscess, coronal pulp chamber, and root canal samples, respectively, after filtering (Table 2). Operational taxonomical units (OTUs), which is an operational definition of a species or group of species often used only when DNA sequence data are available, were utilized as a measure of diversity [17]. In the present study, OTUs were defined at the 3% divergence (97% similarity) threshold using the average neighbor clustering algorithm. The average numbers of OTUs were 165, 199, 185, 56, and 65, respectively. There were 3–4 times more OTUs in the perio group than in the endo group. Within the supragingival plaque and subgingival plaque, the MIM mandibular right first molar (ID 46-1 and 2) showed the lowest number of OTUs and the non-MIM teeth (ID Nor-1 and 2) showed higher than average OTUs.

The α-diversity in the samples was assessed by the Ace, Chao 1, and Shannon diversity indices at a 3% distance range (Table 2). Diversity is an indicator of richness (abundance) and evenness (composition) in the samples. The average Ace index was 185.62, 236.09, 215.64, 78.97, and 111.86, respectively. The average Chao 1 index was 181.92, 230.68, 224.18, 77.72, and 83.9, respectively. The average Shannon index was 3.47, 3.58, 3.24, 2, and 2.2, respectively. Diversity showed a similar pattern to the OTUs between and within the groups.

Table 2. Taxonomic identification and assessment of α-diversity.

Group	Sample ID	Taxonomic Identification		α-Diversity		
		Number of Final Reads	Number of OTUs *	Ace	Chao 1	Shannon
Perio	Nor-1	18092	183	201.84	193.62	3.52
	11-1	19109	155	181.94	184	3.32
	16-1	18517	161	178.61	171.12	3.71
	26-1	18293	196	219.54	206.50	3.67
	36-1	18317	193	212.38	206.04	3.88
	46-1	19395	104	119.42	130.25	2.74
	Average	**18621**	**165**	**185.62**	**181.92**	**3.47**
	Nor-2	19134	206	250.28	243.27	3.70
	11-2	18492	262	319.99	304.78	3.84
	16-2	19032	185	217.45	216.95	3.54
	26-2	18528	201	213.69	207.56	4.01
	36-2	19629	206	238.47	248.50	3.58
	46-2	19491	133	176.67	163.00	2.78
	Average	**19051**	**199**	**236.09**	**230.68**	**3.58**
	46-3	19952	185	215.64	224.18	3.24
Endo	16-4	17033	52	67.61	61.43	2.01
	36-4	19238	59	90.32	94	1.98
	Average	**18136**	**56**	**78.97**	**77.72**	**2.00**
	16-5	16233	62	73.68	69.80	2.23
	36-5	18570	68	150.03	98.00	2.17
	Average	**17402**	**65**	**111.86**	**83.9**	**2.2**

* OTUs: operational taxonomical units.

3.3. Taxonomic Identification of the Oral Microbiome

A total of seven major bacterial phyla (*Spirochaetes, Firmicutes, Fusobacteria, Actinobacteria, Proteobacteria, Bacteriodetes*, and *Saccharibacteria_TM7*) were identified at a higher than 1% frequency in all the plaque samples (Figure 5A). The perio group showed an average of 5.7 phyla. Only sample 36-2 showed seven phyla, including *Spirochaetes*. Seven samples showed six phyla and six samples showed five phyla. The endo group showed an average of 3.5 phyla and sample 36-4 showed two phyla.

Figure 5B shows that at the genus level, *Veillonella, Streptococcus*, and *Leptotrichia* were present at 5% frequency in almost all the perio group samples. *Parvimonas, Peptostreptococcus, Atopobium*, and *Dialister* were present in all endo group samples. Other genera were found in several samples in relation to the habitat or tooth. *Aggregatibacter actinomyces* were not identified.

Figure 5C shows the analysis of the endo group samples at a frequency of 1%. *Bulleidia* was found in all four samples. *Fusobacterium, Mogibacterium, Prevotella*, and *Olsenella* were present in three samples. *HQ441590_g* was found in two samples, while *Shuttleworthia, Alloprevotella, Eubacterium_g10, Selenomonas*, and *Oribacterium* were present in one sample.

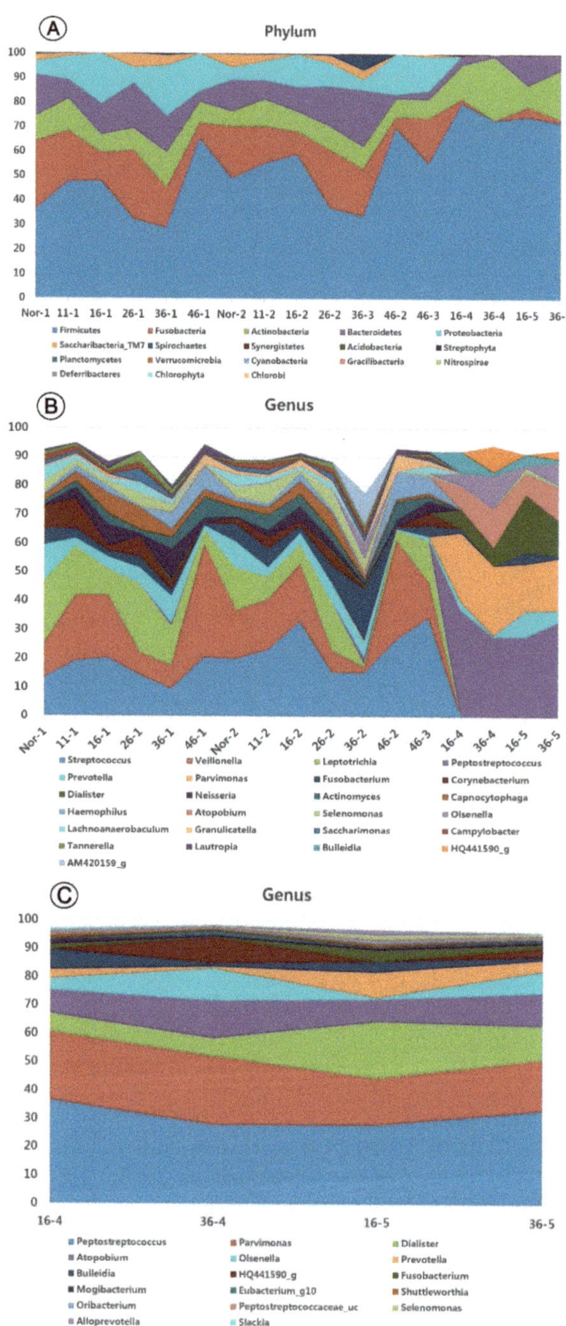

Figure 5. Relative abundance of oral bacteria at the phylum and genus levels. (**A**) Relative abundance of major bacterial phyla with >1% frequency in all samples. (**B**) Relative abundance of major bacterial genera with >5% frequency in all samples. (**C**) Relative abundance of major bacterial genera with >1% frequency in the endodontic samples.

3.4. Heatmap Analysis

In general, there was a distinct difference between the perio and endo groups. The perio group showed genera *Streptococcus, Veillonella, and Leptotrichia* (Figure 6A). Especially, *Fusobacterium* was found in a higher proportion in sample 36-2 compared to that in the other samples of the perio group. The endo group showed significant proportions of gram-positive anaerobic cocci species, such as *Parvimonas micra* and *Peptostreptococcus stomatitis* (group) (Figure 6B).

Figure 6. Heatmap analysis at the genus (A) and species (B) level with >3% frequency. The perio group showed genera *Streptococcus, Veilonella, Leptotrichia, Prevotella,* and *Fusobacterium.* The endo group showed gram-positive anaerobic cocci species, such as *Parvimonas micra* and *Peptostreptococcus stomatitis.*

4. Discussion

The patient showed the stereotypic characteristics of a MIM patient. Based on our study in 2014 [1], the average age of 12 MIM patients at the initial visit was about 7.8 years (range 4–13 years) and it was the same with this patient. This seems to be due to the eruption of the MIM teeth including the permanent first molars. The patient in this study was female and the sex ratio was even in a previous study [1]. However, in the study by Kim et al. [8], there were 24 males out of 38, and in the study by Vargo et al. [3], there was a male predominance at 56.3%. The gender ratio is still controversial.

Interestingly, the patient had a medical history of systemic disease, preterm birth at nine months, and medication for ADHD. Reportedly, almost 95% of the MIM patients had diseases during infancy [8]. One of 12 in Lee's study [1], one of one in McCreedy's 2015 study [18], and four of 30 in Wright's 2016 study [19] were born preterm. In addition, meningitis, spina bifida, cerebral cyst, cephalohematoma, seizures, hydrocephalus, and other diseases were present in MIM patients [3,8,19]. According to a recent review, "the exact aetiology of this condition is unknown. However, the fact that root malformations are limited to isolated teeth suggests that a non-genetic, environmental factor related to past medical history could be the cause" [3]. The authors suggested a possible epigenetic association, which needs further investigation.

In this patient, all the permanent first molars, the deciduous molars, and the maxillary central incisors were MIM teeth. This case seemed more serious compared to those in previous studies [3,8,19]. MIM teeth occurred in the permanent first molar, and often only in the mandible or in the maxilla. In many cases, the primary second molar was affected, but rarely the primary first molar or the permanent second molar. In the future, a classification for MIM is needed. The prevalence of incisors affected by MIM was about half [1] or one-third [3].

Sampling was the most crucial part for identifying the microbiome by collecting plaque and obtaining accurate data. If the sampling is incorrect, the disease-related bacteria may not be accurately identified. In this study, the sampling method applied was the one used by researchers in previous studies [13–15]. The NGS method MiSeq was developed in 2011 by Illumina. The characteristics of MiSeq consists of minimization while maintaining the chemistry of HiSeq (Illumina) and is suitable for reading V3 and V4 of 16S rRNA applied to microbial community analysis [20].

A total of 17 samples were categorized into the perio and endo groups. The perio group had higher average OTUs and alpha diversity compared to the endo group. At the genus level, *Streptococcus*, *Veillonella*, and *Leptotrichia* were present in most of the perio group samples, but not in the endo group. These genera were usually found in healthy Korean preschool children [21].

Particularly, sample 36-2 with a clinically severe dentoalveolar abscess showed only the presence of the *Spirochetes* phylum. *Spirochetes* were present in subgingival plaques and elevated in advanced periodontal disease and endodontic infection [22]. *Treponema socranskii* and *Treponema medium* appeared at 1% frequency in sample 36-2. *Treponema* genera belonging to oral *Spirochetes* were isolated from the periodontal inflammation site [23].

In the heatmap, the perio group was subtly divided into three clusters: (1) 11-2, 11-1, Nor-2, 26-2, 26-1, Nor-1, and 36-1; (2) 16-2, 16-1, 46-3, 46-2, and 46-1; and (3) 36-2. The first group included the MIM-incisor and non-MIM teeth, and the maxillary left MIM-molar. The incidence of *Leptotrichia*, *Prevotella*, and *Fusobacterium* was relatively high. The second group included the right MIM-molars with relatively high *Streptococcus* and *Veillonella*. Sample 36-2 showed a relatively large number of *Fusobacterium*. *Fusobacterium* is an absolute anaerobic gram-negative bacterium with a long filament, which is isolated from periodontal tissue and can aggregate with other oral bacteria [24]. It is one of the signs of acute dentoalveolar infection [15].

Among the endo group, samples 16-4, 16-5, and 36-5 showed four phyla and 36-4 showed two phyla. *Saccharibacteria_TM7* and *Proteobacteria* was not present as compared to the Perio group. At the genus level, *Parvimonas*, *Peptostreptococcus*, *Atopobium*, and *Dialister* were present in all the endo group samples. The anaerobic gram-positive cocci *Parvimonas* and *Peptostreptococcus* have been identified from caries dentin, infected pulp, and complex infections such as dental root canals, progressive periodontitis, and dental abscesses [15,25]. *Atopobium* was originally identified as an anaerobic lactic acid bacillus in a periodontal pocket. *Dialister* is a gram-negative bacterium, which is an absolute anaerobic and found in root canal infections and periodontal infections [25]. In addition, *Shuttleworthia*, *Bulleidia*, *Mogibacterium*, and *Olsenella*, identified at 1% frequency, are present in the root canals in endodontic-periodontal lesion [25]. Regarding species, the endo group showed significant numbers of gram-positive anaerobic cocci species, such as *Peptostreptococcus stomatitis* and *Parvimonas micra*.

According to the new classification of periodontal disease in 2018, general periodontitis is different from MIM complications because they usually start after middle-age [26]. In addition, early-onset periodontitis (EOP), localized juvenile periodontitis (LJP), or aggressive periodontitis (AP), categorized as periodontitis in the 2018 classification, were also different [27]. Although the location in the molar and incisor, and the sudden periodontal abscess were similar to MIM complications, MIM complications develop before puberty while EOP, LJP, and AP develop during puberty. AP was associated with *Aggregatibacter actinomyces*, but it was not identified in the MIM-complicated molar.

In the 2018 classification, "Other conditions affecting the periodontium" included periodontal abscess, endodontic-periodontal lesions, and localized tooth-related factors [28]. In the localized tooth-related factors, tooth anatomy such as cervical enamel projections, enamel pearls, and developmental grooves could enhance plaque retention [29,30]. Due to the inability to keep these areas clean, it was always accompanied by a poor prognosis despite adequate treatment. According to previous studies, the root malformation of MIM resembled cervical enamel projections, enamel pearls, and developmental grooves and cementum deformation were also confirmed [2]. Biofilm was present in the cervical lower part of the crown in the study by Witt et al. [31]. Periodontitis in MIM seemed to begin as localized tooth-related periodontitis and on further progression, it was assumed to be a periodontal abscess with a severe anatomic alteration.

Pulp necrosis occurs in all teeth affected by MIM regardless of the severity of the periodontitis. Pulp necrosis of MIM teeth could spontaneously occur due to the lack of nutrients and oxygen supply [32]. The anatomical characteristics of MIM seem to cause pulp necrosis and localized tooth-related periodontitis independently, and when combined, an endo-perio lesion might form. If a purulent bacterial infection was added to this, a periodontal abscess could occur [32]. There are two case reports of successful endodontic treatment of MIM teeth. Yue and Kim [9] successfully performed endodontic treatment of a MIM mandibular left permanent first molar of a 13-year-old boy and Byun et al. [10] treated MIM-suspected maxillary central incisors endodontically in a 12-year-old boy. Even if pulp treatment was performed, the prognosis is poor because the periodontal disease is difficult to treat. The long-term course of these cases should be determined.

Early diagnosis, the timely extraction of a MIM, and orthodontic treatment might be considered the treatment of choice, as suggested in previous studies [7,8,33–36]. This was because the prognosis for anatomical periodontitis is poor. Although there were two successful cases of endodontic treatment in the previously published papers [9,10], as long as the biofilm causing periodontal disease persists, as in the study by Witt et al. [31], it will cause recurring periodontitis. Rather, extraction of the MIM-molar at the proper time when the secondary molar can move mesially and considering orthodontic treatment may prove to be of long-term benefit to the patient [37].

There were limitations to this research. First, clinical tests including vitality tests, periodontal examination, and a periapical radiographic examination were not done at every visit. Second, microbiological analyses should be done in more MIM patients or more non-MIM patients for comparison.

5. Conclusions

The results of this study confirmed that the dentoalveolar infection of a MIM was different from localized juvenile periodontitis. It seemed that pulp necrosis and localized tooth-related periodontitis occurred spontaneously over time after the eruption of the MIM and developed into severe endo-perio lesions or a periodontal abscess. We carefully suggest that the treatment of choice was the extraction of the MIM at the appropriate time.

Author Contributions: Conceptualization, H.-S.L. and S.-C.C.; Methodology, H.-S.L.; software, H.J.K.; validation, K.L.; formal analysis, O.H.N.; investigation, S.-C.C.; resources, M.-S.K.; data duration, H.-S.L.; writing—original draft preparation, H.-S.L. and S.-C.C.; writing—review and editing, H.-S.L. and S.-C.C.; Supervision, project administration, funding acquisition, H.-S.L. All authors have read and agreed to the published version of the manuscript.

Funding: This research was supported by the Basic Science Research Program of the National Research Foundation of Korea (NRF), funded by the Ministry of Education, Science, and Technology (NRF-2016R1C1B1015005 and 2020R1C1C1006937).

Acknowledgments: We thank to Je Seon Song and Soo-Hyun Kim for stating the MIM research together and Jung-Wook Kim for inspiring us.

Conflicts of Interest: The authors declare no conflict of interest.

References

1. Lee, H.S.; Kim, S.H.; Kim, S.O.; Lee, J.H.; Choi, H.J.; Jung, H.S.; Song, J.S. A new type of dental anomaly: Molar-incisor malformation (MIM). *Oral Surg. Oral Med. Oral Pathol. Oral Radiol.* **2014**, *118*, 101–109. [CrossRef]
2. Lee, H.S.; Kim, S.H.; Kim, S.O.; Choi, B.J.; Cho, S.W.; Park, W.; Song, J.S. Microscopic analysis of molar–incisor malformation. *Oral Surg. Oral Med. Oral Pathol. Oral Radiol.* **2015**, *119*, 544–552. [CrossRef]
3. Vargo, R.J.; Reddy, R.; Da Costa, W.B.; Mugayar, L.R.F.; Islam, M.N.; Potluri, A. Molar-incisor malformation: Eight new cases and a review of the literature. *Int. J. Paediatr. Dent.* **2020**, *30*, 216–224. [CrossRef]
4. Pavlic, A.; Vrecl, M.; Jan, J.; Bizjak, M.; Nemec, A. Case report of a molar-root incisor malformation in a patient with an autoimmune lymphoproliferative syndrome. *BMC Oral Health* **2019**, *19*, 49. [CrossRef]
5. Zschocke, J.; Schossig, A.; Bosshardt, D.D.; Karall, D.; Glueckert, R.; Kapferer-Seebacher, I. Variable expressivity of TCTEX1D2 mutations and a possible pathogenic link of molar-incisor malformation to ciliary dysfunction. *Arch. Oral Biol.* **2017**, *80*, 222–228. [CrossRef]
6. Qari, H.; Kessler, H.; Narayana, N.; Premaraj, S. Symmetric multiquadrant isolated dentin dysplasia (SMIDD), a unique presentation mimicking dentin dysplasia type 1b. *Oral Surg. Oral Med. Oral Pathol. Oral Radiol.* **2017**, *123*, e164–e169. [CrossRef]
7. Kim, M.J.; Song, J.S.; Kim, Y.J.; Kim, J.W.; Jang, K.T.; Hyun, H.K. Clinical Considerations for Dental Management of Children with Molar-Root Incisor Malformations. *J. Clin. Pediatr. Dent.* **2020**, *44*, 55–59. [CrossRef]
8. Kim, J.E.; Hong, J.K.; Yi, W.J.; Heo, M.S.; Lee, S.S.; Choi, S.C.; Huh, K.-H. Clinico-radiologic features of molar-incisor malformation in a case series of 38 patients: A retrospective observational study. *Medicine* **2019**, *98*, e17356. [CrossRef]
9. Yue, W.; Kim, E. Nonsurgical Endodontic Management of a Molar-Incisor Malformation-affected Mandibular First Molar: A Case Report. *J. Endod.* **2016**, *42*, 664–668. [CrossRef]
10. Byun, C.; Kim, C.; Cho, S.; Baek, S.H.; Kim, G.; Kim, S.G.; Kim, S.-Y. Endodontic Treatment of an Anomalous Anterior Tooth with the Aid of a 3-dimensional Printed Physical Tooth Model. *J. Endod.* **2015**, *41*, 961–965. [CrossRef]
11. Marsh, P.D. In Sickness and in Health-What Does the Oral Microbiome Mean to Us? *An Ecological Perspective. Adv. Dent. Res.* **2018**, *29*, 60–65. [CrossRef]
12. Belibasakis, G.N.; Bostanci, N.; Marsh, P.D.; Zaura, E. Applications of the oral microbiome in personalized dentistry. *Arch. Oral Biol.* **2019**, *104*, 7–12. [CrossRef]
13. Lee, H.S.; Lee, J.H.; Kim, S.O.; Song, J.S.; Kim, B.I.; Kim, Y.J. Comparison of the oral microbiome of siblings using next-generation sequencing: A pilot study. *Oral Dis.* **2016**, *22*, 549–556. [CrossRef]
14. Camelo-Castillo, A.J.; Mira, A.; Pico, A.; Nibali, L.; Henderson, B.; Donos, N.; Inmaculada, T. Subgingival microbiota in health compared to periodontitis and the influence of smoking. *Front. Microbiol.* **2015**, *6*, 119. [CrossRef]
15. Santos, A.L.; Siqueira, J.F., Jr.; Rocas, I.N.; Jesus, E.C.; Rosado, A.S.; Tiedje, J.M. Comparing the bacterial diversity of acute and chronic dental root canal infections. *PLoS ONE* **2011**, *6*, e28088. [CrossRef]
16. Hsiao, W.W.; Li, K.L.; Liu, Z.; Jones, C.; Fraser-Liggett, C.M.; Fouad, A.F. Microbial transformation from normal oral microbiota to acute endodontic infections. *BMC Genom.* **2012**, *13*, 345. [CrossRef]
17. Blaxter, M.; Mann, J.; Chapman, T.; Thomas, F.; Whitton, C.; Floyd, R.; Abebe, E. Defining operational taxonomic units using DNA barcode data. *Philos. Trans. R. Soc. Lond. B Biol. Sci.* **2005**, *360*, 1935–1943. [CrossRef]
18. McCreedy, C.; Robbins, H.; Newell, A.; Mallya, S.M. Molar-incisor Malformation: Two Cases of a Newly Described Dental Anomaly. *J. Dent. Child.* **2016**, *83*, 33–37.
19. Wright, J.T.; Curran, A.; Kim, K.J.; Yang, Y.M.; Nam, S.H.; Shin, T.J.; Hyun, H.-K.; Kim, Y.-J.; Lee, S.-H.; Kim, J. Molar root-incisor malformation: Considerations of diverse developmental and etiologic factors. *Oral Surg. Oral Med. Oral Pathol. Oral Radiol.* **2016**, *121*, 164–172. [CrossRef]
20. Qin, N.; Li, D.; Yang, R. Next-generation sequencing technologies and the application in microbiology–A review. *Wei Sheng Wu Xue Bao* **2011**, *51*, 445–457.
21. Lee, S.E.; Nam, O.H.; Lee, H.S.; Choi, S.C. Diversity and homogeneity of oral microbiota in healthy Korean pre-school children using pyrosequencing. *Acta Odontol. Scand.* **2016**, *74*, 335–336. [CrossRef]
22. Sakamoto, M.; Siqueira, J.F., Jr.; Rocas, I.N.; Benno, Y. Diversity of spirochetes in endodontic infections. *J. Clin. Microbiol.* **2009**, *47*, 1352–1357. [CrossRef]
23. Baumgartner, J.C.; Khemaleelakul, S.U.; Xia, T. Identification of spirochetes (treponemes) in endodontic infections. *J. Endod.* **2003**, *29*, 794–797. [CrossRef]
24. Han, Y.W. Fusobacterium nucleatum: A commensal-turned pathogen. *Curr. Opin. Microbiol.* **2015**, *23*, 141–147. [CrossRef]

25. Gomes, B.P.; Berber, V.B.; Kokaras, A.S.; Chen, T.; Paster, B.J. Microbiomes of Endodontic-Periodontal Lesions before and after Chemomechanical Preparation. *J. Endod.* **2015**, *41*, 1975–1984. [CrossRef]
26. Needleman, I.; Garcia, R.; Gkranias, N.; Kirkwood, K.L.; Kocher, T.; Iorio, A.D.; Moreno, F.; Petrie, A. Mean annual attachment, bone level, and tooth loss: A systematic review. *J. Periodontol.* **2018**, *89* (Suppl. S1), S120–S139. [CrossRef] [PubMed]
27. Fine, D.H.; Patil, A.G.; Loos, B.G. Classification and diagnosis of aggressive periodontitis. *J. Periodontol.* **2018**, *89* (Suppl. S1), S103–S119. [CrossRef]
28. Caton, J.G.; Armitage, G.; Berglundh, T.; Chapple, I.L.C.; Jepsen, S.; Kornman, K.S.; Mealey, B.L.; Papapanou, P.N.; Sanz, M.; Tonetti, M.S. A new classification scheme for periodontal and peri-implant diseases and conditions—Introduction and key changes from the 1999 classification. *J. Periodontol.* **2018**, *89* (Suppl. S1), S1–S8. [CrossRef]
29. Ercoli, C.; Caton, J.G. Dental prostheses and tooth-related factors. *J. Periodontol.* **2018**, *89* (Suppl. S1), S223–S236. [CrossRef]
30. Romeo, U.; Palaia, G.; Botti, R.; Nardi, A.; Del Vecchio, A.; Tenore, G.; Polimeni, A. Enamel pearls as a predisposing factor to localized periodontitis. *Quintessence Int.* **2011**, *42*, 69–71.
31. Witt, C.V.; Hirt, T.; Rutz, G.; Luder, H.U. Root malformation associated with a cervical mineralized diaphragm–a distinct form of tooth abnormality? *Oral Surg. Oral Med. Oral Pathol. Oral Radiol.* **2014**, *117*, e311–e319. [CrossRef]
32. Herrera, D.; Retamal-Valdes, B.; Alonso, B.; Feres, M. Acute periodontal lesions (periodontal abscesses and necrotizing periodontal diseases) and endo-periodontal lesions. *J. Periodontol.* **2018**, *89* (Suppl. S1), S85–S102. [CrossRef]
33. Neo, H.L.; Watt, E.N.; Acharya, P. Molar-incisor malformation: A case report and clinical considerations. *J. Orthod.* **2019**, *46*, 343–348. [CrossRef]
34. Yoshiaki, N.; Erika, K.; Ayako, O.; Ryoko, O.; Mieko, S.; Yasuko, T.; Chieko, T.; Kazumune, A.; Hideki, D.; Tamotsu, S.; et al. Oral micobiome in Four Female Centenarians. *Appl. Sci.* **2020**, *10*, 5312.
35. Yoshiaki, N.; Erika, K.; Noboru, K.; kaname, N.; Akihiro, Y.; Nobuhiro, H. The Oral Microbiome of Healthy Japanese People at the Age of 90. *Appl. Sci.* **2020**, *10*, 6450.
36. Hwang, J.Y.; Lee, H.-S.; Choi, J.; Nam, O.H.; Kim, M.S.; Choi, S.C. The Oral Microbiome in Children with Black Stained Tooth. *Appl. Sci.* **2020**, *10*, 8054. [CrossRef]
37. Cervino, G.; Cicciù, M.; Biondi, A.; Bocchieri, S.; Herford, A.S.; Laino, L.; Fiorillo, L. Antibiotic Prophylaxis on Third Molar Extraction: Systematic Review of Recent Data. *Antibiotics* **2019**, *8*, 53. [CrossRef]

MDPI

St. Alban-Anlage 66

4052 Basel

Switzerland

Tel. +41 61 683 77 34

Fax +41 61 302 89 18

www.mdpi.com

Applied Sciences Editorial Office

E-mail: applsci@mdpi.com

www.mdpi.com/journal/applsci